MORE THAN HONEY

MORE THAN HONEY

The Survival of Bees
and the Future of Our World

MARKUS IMHOOF & CLAUS-PETER LIECKFELD

Photographs by Hans-Jürgen Koch & Heidi Reschke-Koch

DAVID SUZUKI INSTITUTE

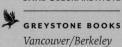

GREYSTONE BOOKS
Vancouver/Berkeley

Greystone Books Ltd.
www.greystonebooks.com

David Suzuki Institute
219–2211 West 4th Avenue
Vancouver BC Canada V6K 4S2

Cataloguing data available from Library and Archives Canada
ISBN 978-1-77164-099-2 (pbk.)
ISBN 978-1-77164-100-5 (epub)
ISBN 978-1-77164-101-2 (epdf)

Editing by Lesley Cameron
Cover and text design by Nayeli Jimenez
Photographs by Hans-Jürgen Koch and Heidi Reschke-Koch
Printed and bound in China
by 1010 Printing International Ltd.
Distributed in the U.S. by Publishers Group West

We gratefully acknowledge the financial support of the Canada Council for the Arts, the British Columbia Arts Council, the Province of British Columbia through the Book Publishing Tax Credit, and the Government of Canada through the Canada Book Fund for our publishing activities.

Greystone Books is committed to reducing the consumption of old-growth forests in the books it publishes. This book is one step toward that goal.

Contents

PREFACE
Man Shall Not Live by Bread Alone   *vii*

1   Bees: A Big Business   *1*

2   Bees in an Ideal World   *21*

3   Bees in the Lab   *39*

4   Customized Bees   *59*

5   Humans as Bees   *79*

6   Bees: An Untamed Force   *95*

7   Bees of the Future   *113*

8   The Origins of the
Documentary *More Than Honey*   *127*

NOTES   *149*
REFERENCES   *153*
INDEX   *156*
ACKNOWLEDGMENTS   *161*

Preface

MAN SHALL NOT LIVE BY BREAD ALONE

"A bee colony is like a magic well;
the more you draw from it the richer it flows."
KARL VON FRISCH

UNTIL RECENTLY, THE May 24, 2012, article in the *Neue Zürcher Zeitung* would not have been considered front-page news for this well-respected Swiss newspaper: "Bee Smuggling Exposed: 80 Illegally Imported Bee Colonies Destroyed." A German man aged almost thirty was arrested by customs investigators as he attempted to sell artificial swarms—bees taken by the kilo from other colonies with a separately packed queen—that had not been declared and were thus illegal. Among the twenty Swiss beekeepers he had contacted online and lured to the border was an undercover buyer from the Verein deutschschweizerischer Bienenfreunde (the society of German and Swiss bee friends). He revealed his true intentions just as the money was being exchanged for bees. All of a sudden it wasn't bees that were swarming but customs officials. This single operation brought the Zurich customs investigators eighty Swiss buyers whose illegally imported colonies were immediately and completely destroyed. Many more buyers remained at large.

Why were the colonies destroyed? Why so much fuss about a mere 135 Swiss francs (slightly more than US$150) per colony? For once, the customs investigators were not interested in money, their motive was the survival of the Swiss bee world. Illegally imported colonies could be disease carriers, worsening the rates of bee mortality there. In some regions of Switzerland, up to 70 percent of colonies didn't survive the winter of 2011/2012. In Germany, according to expert estimates, up to one-third of the colonies—about 300,000 from a total of roughly 1 million—didn't survive the winter.[1] And in the USA, on average, one-third of all bee colonies have been dying annually since 2006. At the fifth EurBee Congress, an international conference of apidology, in Halle, Germany, at the beginning of September 2012, Professor Robin Moritz warned about the worldwide collapse of bee populations in his opening address.

If 70 percent of all cattle or 30 percent of all chickens were to die annually, states of emergency would be declared everywhere. The death of bees is at least that dramatic and with even more far-reaching consequences. The bee is our smallest working animal. In the peak year of 2007, they gave us a record yield worldwide of 1.4 million metric tons (slightly more than 1.5 million short tons) of honey. But the questions arising from the consequences of these deaths are about more than just honey.

Successful pollination is a prerequisite for fruit production. Bees pollinate more than 70 percent of the one hundred most important domesticated plant species in the world. A colony of honeybees can visit up to 7 million blossoms every day. It is hard to imagine what a loss of bee power would mean. What if the honeybees, whose services humans have taken advantage of for years, no longer performed their services? Bees are responsible for 30 percent of global harvests, and if they fail, we have to do without every third bite. It is about nothing less than global food supplies. Our plates would look dreary were it not for the bees' contribution. Many of the things that are colorful, aromatic, and tempting would be missing. Apples, cherries, asparagus, soya beans, peaches, and cucumbers, for example, are only a few of almost one hundred kinds of fruit and vegetables that are dependent on pollination by bees. A hamburger would have no salad, no onion, no ketchup; the meat would come from a cow that had never eaten clover. Only the bread roll would be unaffected as wheat is pollinated by the wind.

In the quest to explain why honeybees are leaving us by the billions, we find not a reason but rather reasons: diseases, including epidemics; agricultural toxins; depletion of blossoms leading to starvation for bees; changes in climate conditions; and a weakening of the natural resistance of bees. The majority of experts believe that it is the sum of many different, intensifying attacks on bees' immune systems that is now causing catastrophic gaps in the global bee population.

Many experts and apiarists place the blame primarily on pesticides. Scientists have been able to prove that in 2008 in the Rhine Valley, more than eleven thousand bee colonies were either killed or severely damaged by a nicotine-like neurotoxin used when sowing corn.

On March 10, 2011, the United Nations (UN) issued a statement in response to the crisis: "Systemic insecticides such as those used as seed coatings, which migrate from the roots through the entire plant all the way to the flowers, can potentially cause toxic chronic exposure to non-target pollinators. [...] Laboratory studies have shown that such chemicals can cause losses of sense of direction, impair memory and brain metabolism, and cause mortality."[2]

The UN has stated in at least one of its World Food Reports that the world's population can only be supported by small, structured farming. However, in reality the opposite continues to be the case because monocultures are more efficient to farm. Just as totalitarian systems can only survive with a brutal police force, monocultures rely on policing by pesticides that keep in check pests that would otherwise find their ideal living conditions there. Toxins in foodstuffs and the loss of bees are the collateral damage. Intensive farming methods for agricultural rationalization and improved efficiency are justified by the claim that there is no other way of guaranteeing global nourishment. "Humans have the illusion that in the twenty-first century they can be independent of nature with the aid of technological progress," said Achim Steiner, director of the UN Environment Programme.[3] This raises a loaded question: Do we want to starve healthily or to eat but be poisoned in the process?

However, some experts from other renowned bee institutes don't blame agrochemicals for the decline in bees. Dr. Peter Rosenkranz, from the Landesanstalt für Bienenkunde of Hohenheim University, is one of them: "The most important factor is *Varroa destructor*, followed by *Varroa destructor* and then *Varroa destructor*."[4]

Varroa destructor, or the *varroa* mite, was introduced from China and has unequivocally been causing honeybees problems since the end of the 1970s. These mites infest the broods and live off the blood of bees. Additionally, viruses that deform the wings infiltrate the open wounds at the point of contact of the bites. In human terms, this mite would correspond to a leech the size of a rabbit. The immense deficiencies within populations lead to labor shortages and thus to a lack of care for the broods and insufficient food reserves. The colony becomes weak and eventually collapses. Does this collapse mean the end of the mites too? Unfortunately not. Weakened colonies are plundered by stronger ones—after all, where else can nectar be more easily foraged than where the stocks have already been fully processed?—and the raiders are in turn jumped on by the *Varroa destructors*, which hitch a ride with them to other, possibly uninfected hives. Furthermore, there are around a dozen serious diseases and parasites affecting honeybees that beekeepers keep under control by using chemicals that are to some extent harmful.

Can all these factors combine to explain the phenomenon that has become famous under the name of Colony Collapse Disorder (CCD)? Beekeepers, particularly in the USA but also in Europe, have come to accept that in some years their colonies simply disappear without a trace. Only the queen and a few bees remain on the honeycomb, which is well filled with honey and pollen. All the others are gone, and no nurse bees remain to feed the abandoned brood. There are no bodies of bees in the hives, none near the hive opening, and none nearby. They have disappeared, apparently without any cause, without previously displaying any symptoms of disease, and without showing up anywhere else.

On the basis of this dramatic development, a growing number of scientists worldwide are becoming interested in the problems of the honeybee, both researching the causes and attempting to find solutions. Some beekeepers are also seeking alternatives to conventional beekeeping in order to focus on their natural living habits and reproductive practices. There is hope that in this way bee colonies can be invigorated and then use their own strengths to better come to terms with damaging environmental factors that can affect tended bees which are completely dependent on human beings for support. After all, bees have survived on this planet without human support for at least 80 million years.

Over this time they have not only adapted to continuous changes in their environment but also perfected the art of coexisting with humans—without whom they successfully survived for centuries. Experts believe that the oldest preserved specimen found in amber belonged to the eusocial variety of bees. The eusocial bee is strictly organized, disciplined, and efficient—it's not surprising that humans quickly realized what an excellent partner this insect could be. Unlike a farm horse or load-bearing elephants, bees are not forced into doing something that they wouldn't usually be doing. Everything from which we profit—the hauling of sweetness back to the hive and creating the preconditions for a successful harvest—they do better than we could ever teach them to. All we humans have to do is monitor an existing, functioning system. But now it is becoming apparent that human intrusions in the functioning system have destabilized it.

In the USA, one-sixth of all bee colonies stem from only 308 queen bees, which has led to a massive depletion in genetic diversity. Worldwide, the flagging situation of the honeybees can also be traced to the fact that the gene pool is continually being reduced. The focus on breeding bees that tend to industriousness and gentleness has come at the expense of their health, their ability to live.

Like the documentary *More Than Honey*, this book of the same name follows people who live with, for, and from bees: large-scale operators, beekeepers, breeders, and scientists. It looks at those who transport their bees over thousands of kilometers across the continent, those mailing them throughout the world, those wanting to protect bees through racial purity, those trying to look into their brains, those wanting to replace them with their own labor. As diverse and, to some extent, absurd as the respective approaches seem, all the participants have a common love of bees—and still something is wrong. It would be disastrous if the relationship between humans and bees over thousands of years became a war between civilization and nature.

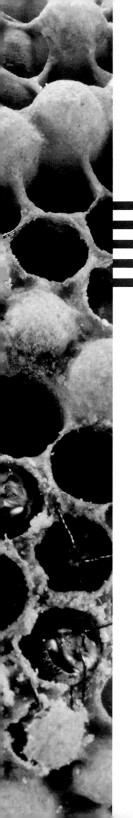

BEES: A BIG BUSINESS

ONCE UPON A time there was a beekeeper in tiny, wintery, Blackfoot, Idaho, who thought long and hard about how he could best use the long, work-free winter months. He liked the idea that in warmer climes the flight and working times of his bees would be naturally longer. Maybe, he thought, the bees could be transported before the cold snap to distant but warm California, where they could continue to gather honey beyond their normal yearly schedule. That was in 1894, and the man, Nephi Ephraim Miller, is today considered to be the pioneer of large-scale US migratory beekeeping. After a successful trial run, in 1895 he packed his colonies onto trains to California, and watched with satisfaction as they yielded good returns at a time of the year when at home they would have still been half asleep and waiting for the onset of spring.

It would have certainly given N.E.—the abbreviation by which he is known to all who are interested in the history of big business with bees in the USA— even more satisfaction to know that his great-grandson John followed in his footsteps. What is more, John has turned his pioneering great-grandfather's road into a highway. But we cannot give an account of this without including an account of the Californian almond empire.

Eighty percent of all almonds eaten globally—be they raw or roasted, ground or processed into marzipan—are harvested in California. In the trading year 2011/2012, the Golden State exported more than 453 million kilograms (around 1 billion pounds) of them. China, Spain, Germany, India, and the United Arab Emirates are the biggest importers, accounting for 53 percent of the total market. More than 76 million kilograms (167 million pounds) go to China, leaving around 222 million kilograms (490 million pounds) for the US domestic market.

In order to harvest such immense amounts, correspondingly huge num-bers of plants have to be pollinated. Three-quarters of all bee colonies in the USA, around 93 billion individual bees, are out and about for just over four weeks in February and the beginning of March, pollinating almond blossom in an area covering roughly three thousand square kilometers (approximately 1,158 square miles).

An armada of trucks is on the road to drive them in swarms overland in a migration scheduled and executed by humans. The average US bee needs to be fit not only for normal bee work but especially so for the stressful journey along the almost endless highways. Some migratory beekeepers from Florida fly their bees to California in February for the almond blossom, then up to Washington State for apple and cherry pollination, then all the way back to Florida for citrus fertilization, before sending them to New England for the blueberries, and finally back to Florida for the winter.

One of the largest pollination contractors and honey producers in the USA is N. E. Miller's great-grandson John, born in 1954. Honey rhymes with money, a coincidence immediately noticeable as the super-fit marathon runner from the small North Dakota town of Gackle, with a population of just over 300, explains his business. Miller is someone who loves his bees, his *dancing ladies* as he calls them. Nevertheless, he exploits them. He lives with this paradox because his business would not function otherwise; massive intrusions into animal life are the inevitable collateral damage of his success.

The Miller bees that overwinter under permanent temperature control in potato stores in Idaho are woken from their winter dormancy two weeks before the almonds blossom. In January, the majority of the fifteen thousand colonies are freighted by trucks from their winter depot in Idaho via the Donner Pass to California for pollination duties, a journey that often requires the drivers to fit snow chains. From the end of January, Miller and his team distribute the colonies over a 320-kilometer (two-hundred-mile) stretch between Modesto and Chico among the endless rows of almond trees—the trees have no leaves, as they grow after the blossom. Then it's just a question of waiting until the blossom appears. In the meantime, the bees are fed sugar water.

When the time comes, the pollination operation begins. Normally, the bees would undertake an orientation flight in a new location. But John Miller's bees don't have any choices as they are surrounded by almond trees. As far as the eye can see, there are no other blossoms to be seen.

From the middle of February, the undulating landscape is blossom pink. On March 1, the almond blossoms are reaching their peak, and a good two weeks later, the "pollination guys" quickly begin to gather their colonies and prepare them for departure. The almond plantations are extreme monocultures, everything else that blossoms and could provide nutrition to bees is meticulously eradicated—simply so that the bees are not distracted by other blossoms. "They [the pollination guys] should do what they are paid to do, pollinate almond blossom," says Miller, who pockets US$150 per colony for the almond operation. With 15,000 colonies, it costs US$2.25 million for the month-long pollination services.

For the almond barons of the West Coast, that's 28 percent of their total production costs. That may sound a lot, but there is no other workforce, no machine that could do this work for comparable costs. Only with the honeybee can such well-organized armies of temporary workers be so effectively deployed. Miller then makes one of those remarks that make him *America's favorite bee guy*, at least for the media and as a talk show guest: "Pollination is a whore's job: I arrive at night, wearing a veil, they pay me, a couple of weeks later they call me and tell me kindly to beat it."

When the colonies have done their work and are again transported from California, their challenge is to survive the heat. "If the truck gets stuck in a jam or breaks down with engine problems, you can cook a truckload of bees to death in two hours," said Miller.

The truckers stop only occasionally, and then mostly just to spray the tarpaulins under which the cases with the bees are firmly secured. The drivers drink as little as possible to reduce the need for restroom breaks. Every journey is a race against death. Limiting travel time to the cool night hours would be good for the bees but not for the schedule, and so the animals are overtaxed for the whole time on the road, speeding along the highway under the blazing sun. The colonies can only survive this if there is enough airflow through the netting from the cruising speed. It is a tough job for both the truckers and the bees. But as the *bee guys* say, they have no choice.

Not only that, but even here, far away from the huge monocultures, the bees are subject to threats, the three Ps: pesticides, parasites, and pasture loss.

The Mexican migrant workers who do the spraying in the almond plantations also have no choice. The sprays used are mostly fungicides and fungistatics, and without them the gigantic Californian monoculture would immediately buckle. As with all monocultures, harmful organisms exponentially increase if their hosts are available in almost unlimited quantities.

Signs on the access roads draw attention to the fact that those who enter the plantations accept an increased risk of cancer. The warnings are not for the protection of the workers—most of the Mexicans cannot speak English—but to prevent possible claims for damages. In almond plantations, as with everywhere else where large-scale agriculture is practiced, whether it's cramming animals into stalls or plants into fields, nothing happens without the assistance of agrochemicals.

Workers have to spray when the blossoms are open, which means during the day when the bees are busy. From the perspective of yields, the fact that the gatherers collect nectar with agents that kill off fungi and spores plays a secondary role at most. Almond blossom honey is as good as inedible and is left in the honeycomb; "only" the brood is harmed and even killed by the fungicide honey. But in the almond business, it is more economical to factor in the losses to the beekeeper than to do without the brutal, bee-harming spraying operations. Miller grudgingly admits that "it's a pact with the Devil."

Miller's year is a tightly scheduled series of transports, bee assignments, splitting, short breaks, and hibernation. During the short winter break, he negotiates the terms of business for the coming year's operations. Bee brokers mediate between the prospective customers—for example, the almond producers of California or the peach farmers of Georgia—and the suppliers, the beekeepers throughout the USA. There is a great deal of money involved. "I hear the sound of money," said Miller as he released his bees in the almond plantation. The sound of this money is the buzz and hum that only bees can make.

At the beginning of the pollination period, Miller usually meets up with his friend MacIlvaine, an almond farmer with whom he has worked for many years, in the seemingly endless rows of trees. The *almond guys*, and even more so the *bee guys*, are men with conservative values. Theirs are not proper friendships but trusted business partnerships that will not be dropped just because someone else, somewhere else offers a few dollars more. And many beekeepers are not simply conservative in their values, they are generally arch-conservative. John Miller doesn't belong to this group, but he did get very agitated about the "Beekeepers for Obama" initiative. A *bee guy* is a Republican, period! This did not stop him from suggesting to the Democrat Obama that beehives be put in the grounds of the White House, though.

Miller is big in the migratory bee business but he is not the biggest. His colleague Richard Adee, from neighboring South Dakota, has 23,000 colonies. Like Miller, he moves them along the highways. When the almond blossom season is ending, some of Adee's trucks roll from the southwest of the USA way up north to the Canadian border to produce honey. The other bees that have survived almond pollination remain in California to be "split." Splitting is what beekeepers around the world do when two or even more colonies are to be made from one. There are a variety of ways to approach splitting, and the idea of splitting using a conveyor belt came to Miller while he was training on a treadmill at a fitness studio.

The operation takes place on a scale of industrial farming and places the greatest imaginable stress on a bee colony. The beehives are placed on a conveyor belt, and automatic brushes sweep the bees away while a blade scrapes off the wax so that the honeycomb frames can be more easily extracted. Then migrant workers wearing protective clothing and looking like the Michelin Man separate the frames and sort them into categories: frames with eggs, frames with worker broods, and frames holding the stocks of nectar and pollen. Other workers sort the frames into new hives so that there is something of everything in each hive. Empty frames are then added, the idea being that the hives will be repopulated by the bees. The bees are then poured over them by the bucketful. These are mostly bees that had sought safety from the scrapers in clusters under the roofing or in nearby trees.

So, one colony becomes four in next to no time. Such a new artificial colony only becomes able to survive when a yet-to-be-fertilized queen is implanted after about three days. Miller buys these from a breeder. The old queen generally doesn't survive the splitting process. If by chance it does survive and lands in one of the four clusters of bees, it fights with the new queen until only one of them is left.

This violent form of population expansion—comparable to ripping open the house of a large family, grabbing the family members, and indiscriminately placing them in new homes and expecting them to carry on their lives with the new family—prevents the bees' impulse to swarm. After their torture, Miller allows them a short period of recovery among the Californian mountain flowers where the new queen can make its nuptial flight. As soon as the queen lays its eggs, the colony has to make the 2,700-kilometer (almost 1,680-mile) trip back to Gackle, North Dakota, where they gather honey for a brand called Dutch Gold Honey. A business partner of Miller is the ex-cycling pro

and current triathlon athlete Lance Armstrong, who ironically enough campaigns for healthy doping with honey energizers when not having to make statements against doping accusations of a completely different kind. Honey production in the summer months is, as it were, a second job of bees; the big dollars come from their main job as pollinators.

In his hometown, Miller is neither the high-flyer of the US bee business nor the TV talk show guest who makes pointed remarks about bees and the beekeeping world, but a normal citizen, albeit an esteemed one, and a philanthropist promoting youth sport and a lively neighborhood. Miller calls the countryside around Gackle "widow land"; most of it belongs to the widows of farmers, the majority of whom have German names: Bechtle, Bader, Müller, Dewald, Kaiser, and Schüler. Their ancestors brought with them a number of generations ago beekeeping skills and a love of bees from their European homelands. Placing his colonies among "the Germans" seems an obvious choice to Miller, but because of the vastness of the country it is not without its complications. In fact, it is a logistical nightmare.

Not only that, but even here, far away from the huge monocultures, the bees are subject to threats, the three Ps: pesticides, parasites, and pasture loss.

Most US citizens accept the first P, pesticides, and their side effects as God-given, akin to the force of gravity or capitalism. In a country where cultivated areas reach to the horizons, farming without agrochemicals is deemed to be almost as impossible as farming without sun and water. Big is still best—and big is only possible with the help of chemicals. That may well be disastrous and ruinous in the medium-term. For the time being, as far as pesticides are concerned, the motto is "the more, the better."

The second P, parasites, is mostly about the *Varroa destructor*, a mite affecting and seriously threatening bee populations around the world. They have to be, and are being, combated—when possible using harmless measures, and in emergencies by whatever it takes. Whether the enemy is called *Varroa* or *Nosema*, a highly infectious global disease, Miller's course of action is antibiotics: "I fight back with all I have ... chemicals!" he says, sounding optimistic. He is careful to stay abreast of new developments and to keep up-to-date with research, an increasing amount of which is looking at the male *Varroa* mites. All in all, he considers himself to be on the right path. The considerably smaller male mites die when they come into contact with natural acids or chemical agents, unlike their tougher female counterparts. Miller is confident that the mighty mite will be beaten in the foreseeable future, an optimistic expectation shared by few others.

Miller finds the third P, pasture loss, almost more worrying than pesticides and parasites. Ironically, the pastureland rich in flowers ideal for bees—literally the land flowing with milk and honey—is shrinking everywhere, including in his native North Dakota. It is rapidly losing ground to soya and grain cultivation. The production of Miller's highly praised light-colored honey—in the USA, in contrast to the rest of the world, light-colored honey is considered the only good honey—is endangered if the bees can only fan out in agricultural wastelands with monocultures as far as the eye can see.

There is another threat, and one that possibly surpasses the dangers of the three Ps. Miller has often experienced and suffered from it, but the shock of his first experience runs deep. One February, Miller had transported his then fourteen thousand beehives from their safe winter home in the former potato stores in Idaho to California and unloaded them at his second home in Newcastle, as he did every year. From there they would be distributed to collection points throughout California Central Valley, sub-centers for distribution in the plantations. But then Miller made a horrifying discovery: most of his bees were gone. Forty percent of his stock, some 150 million individuals, had simply disappeared. Only the queen and a few, clearly bewildered workers were to be found on combs filled with honey. "It was spooky," said Miller.

It also had a name: Colony Collapse Disorder (CCD), to which the international media quickly added the adjective "mysterious," because until now there has been no satisfactory explanation for it, particularly as no dead bees have ever been found near the scene. If we think about it in human terms, it is like a scenario for an apocalyptic movie: The supermarkets and refrigerators are full, babies are crying in the maternity wards, but all the adolescents and adults have vanished from the face of the Earth.

For a while it was thought that the phenomenon could be traced back to 2007, and at that time it attracted a lot of attention. However, US beekeepers had been affected by it earlier, but the reports had not reached the general public, be it from shame or worry that the total loss of their stocks might be interpreted as sloppy beekeeping. Instead of CCD, there was talk of PPB—piss poor beekeeper—another of Miller's quips. PPB means fellow beekeepers who through lack of care only had themselves to blame for losses. Initially, numerous US beekeepers hushed up the extent of their losses in order to avoid any scandal, and this led to delays in searching for the causes. Only when it began to affect many of the big beekeepers was there something like a nationwide "coming out" and the way cleared for a systematic search for the causes.

In 2007 alone, thirty-six US states lost more than one-third of their bee populations. Miller estimates that of the roughly five thousand commercial beekeepers active at the end of the 1990s, only about 1,250 remained in 2012.

(Miller considers a commercial beekeeper to be someone with more than one hundred colonies producing an annual yield of 3,000 kilograms (more than 6,600 pounds).)

And the problem wasn't unique to the USA. All of a sudden CCD was a global problem. Even insularity offered no protection from the evil: In April 2007, Taiwan TV reported that 10 million bees had disappeared without trace. Almost at the same time there were reports from the UK and Canada, where the winter loss rates suddenly doubled—culminating in a total loss of almost one-third of stocks. In New Brunswick, almost 60 percent of the bees completely vanished. All the reports of loss ended—more or less plainly—with the admission that they were groping in the dark as far as the cause of the phenomenon was concerned.

Bee research centers and research units working in the commercial sector all over the world suddenly had a flood of assignments and research findings to record about the insects that they had previously struggled in vain to help.

The research also led to results. There were answers, but not the answer. Scientists from Pennsylvania State University believed that CCD could be a kind of "bee AIDS," a breakdown of the immune system as a result of diverse long-term assaults—attacks in which the *Varroa* mites were possibly only responsible for the final blow. "The magnitude of infectious materials detected in adult males indicates an impairment of the immune system."[1]

A combination of deformed wings, the most striking symptom of *Varroa* attacks, and a yet to be identified infection could have led to the remarkable speed with which CCD spread, according to the tentative explanations of the Pennsylvania Study Group. Viruses, pathogens that were normally unable to penetrate the chitinous exoskeleton of bees, could invade through open wounds. Since then, the situation has become similar to that of HIV/AIDS and cancer research; every three months there are reports of a breakthrough.

At the beginning of 2012, a publication from San Francisco State University surprised the scientific community and raised hopes by announcing that there was maybe one cause or at least one main cause. The potential cause turned out to be a fly, *Apocephalus borelias*, which attaches its eggs to bees. This seemed to rob the bees of their sense of direction, although how and why it did this was still unknown.

The discovery was more or less accidental. An insect expert named John Hafenik collected from the university campus some dead and obviously dying bees whose bodies had been infested by a then unobtrusive fly. The bees, still living but affected by the flies, scrabble around, reeling as if drunk. The phenomenon began to interest the scientist. "Now we have to discover," said Hafenik in early January 2012, "how the parasite influences the behavior of bees." Hafenik thought that the internal clock of the bees, and with it their circadian rhythms, could have been disrupted by the parasites.[2]

Agrotoxins—both individual preparations but also and especially cocktails of more than one type—were high on the list of suspects on the bee battlefield. For farmers, it is more economical to spray compounds than to repeatedly spray their fields with a single chemical. It is difficult to work out which toxins are involved, let alone name the guilty substances, as it is not easy to isolate the various active agents in a mixture. The diversity of the contents of the suspect substances hinders their being singled out.

But in one case at least, there were growing indications of which toxin was to blame. Insecticides such as Imidacloprid and Clothianidin from Bayer have been on the market since 1991. Classed as neonicotinoids, they not only kill the insects they are supposed to, they also harm bees. When all goes well, neonicotinoids can protect sugar beets, Swiss chard, onions, potatoes, corn, and rape against all sorts of pests like plant lice, wireworms, Colorado beetles, cabbage stem flea beetles, and corn borers. Because the active agents are held in the plant bodies and are broken down slowly, the toxic effect remains for a relatively long time. Insects that eat the parts of the plants that have been impregnated with

Insects that eat the parts of the plants that have been impregnated with pesticides die because the substances bring the chemical transmission of signals in the brain to a standstill.

pesticides die because the substances bring the chemical transmission of signals in the brain to a standstill. Unfortunately, the agents do not distinguish
between pests and beneficial insects.

In the 1990s, there were already protests against these substances. In 2003,
French research scientists established that seeds treated with the insecticide
sold under the trade name of Gaucho were lethal for bees. A year later, the then
minister for the environment banned the substance, although it had already
been prohibited for use on sunflowers since 1999.

But the research findings and the ensuing action taken by the French
authorities were not enough to protect the neighboring beekeepers east of the
Rhine. In May 2008, in Ortenaukreis (a district in southern Germany roughly
adjacent to Strasbourg), for example, there was a disastrous mass mortality
of at least 11,500 colonies (some 330 million insects). Some seven hundred
beekeepers lost all or some of their colonies. Detailed studies from the Julius
Kühn-Institut, the German federal research institute for cultivated plants,
revealed that there were "clear indications that Clothianidin was responsible
for the death of bees particularly in parts of Baden-Württemberg."[3]

The sequence of events, too, could be reconstructed. During the sowing
of corn seeds that had been treated with Clothianidin, toxic dust had been
released and consequently contaminated fruit blossoms as well as rape and
dandelions over a wide area. On top of this, unsuitable sowing drills that used
too much compressed air had been used.

The question of who was responsible for the chemical disaster was never
resolved. The company (Bayer) offered no official admission of guilt but did
pay into a compensation fund; the state of Baden-Württemberg compensated
beekeepers from the Upper Rhine on the condition that they did not pur-
sue further claims. In the meantime, the company's claims that their treated
corn seeds disappear harmlessly into the earth where bees are not known to
fly have been refuted. On the one hand, it is unavoidable that some of the
treated seeds should remain on the surface contaminating dewdrops and pud-
dles, typically during the turning maneuvers of seed drills; on the other hand,
heavy rainfall, an increasingly common occurrence in the course of climate

change, results in seeds being washed free. Bees, however, do not have sensors telling them which water is pure and which is contaminated.

Another path of infection is possibly more serious than contaminated water. The seeds of plants that have been treated with neonicotinoids excrete traces of it in their condensation. Bees collect not only nectar and pollen but also water, and there is hardly a more convenient way for it to be consumed than in the form of dewdrops.

We now know that even tiny amounts of neonicotinoids are enough to make bees and bumblebees lose their sense of direction. Worker bees fly off, which in the worst case indirectly kills the colony, according to Mickaël Henry from the Institut national de la recherche agronomique (INRA, the national institute for agricultural research) in Avignon, France. Henry administered small doses of neonicotinoids to test bees—well below the amounts that bees normally absorb—and then fitted them with tiny chips to enable his team to follow their flight. The marked bees returned to the hive two to three times less frequently than those from an untreated control group did. The neonicotinoid-dosed bees got lost, which meant that they died. A honeybee is as likely to survive alone as a baby or an isolated human organ.

According to the April 2012 edition of the scientific journal *Science,* Penelope Whitehorn's research group from Stirling University in Scotland had similar findings. Their test animals were bumblebees that the researchers had fed with rape pollen and nectar, each containing neonicotinoids in amounts similar to those found on sprayed fields. A control group received food without any toxins at all. Both groups were released then recaptured six weeks later and studied. Bees from the group that were given the chemicals were 12 percent lighter and produced about 85 percent fewer queen bees from their ranks. As with all kinds of bees, the survival of a bumblebee colony depends on the fertility of the queen. Both the Scottish and the French research groups recommended "a re-evaluation of the use of neonicotinoids on blossoming plants," wrote the science journalist Jürgen Langenbach.[4]

In Switzerland, where, depending on the canton, 30 percent to 70 percent of the bee colonies did not survive the winter of 2011/2012, "about four [metric] tons of neonicotinoids are used annually on agricultural crops. That is, admittedly, little when compared to the 2,000 [metric] tons of pesticides that land annually on [Swiss] fields but with the neonicotinoids it evidently doesn't depend on the amounts."[5] Even a dosage corresponding to a four-billionth of the bee's body weight is fatal to the organism.[6]

What the critics and aggrieved beekeepers have long claimed—that the active agents in the sprayed substances really do have a direct influence on the demise of bees, even if the correct seed drills are in use—was substantiated, most recently at the beginning of March 2012, with the publication of studies from Harvard University.

The US scientists exposed bee colonies to agents in distinctly lower concentrations than those confronted by bees in their natural habitats. Despite this, 94 percent of all the lab bees died within a short period. One of the reasons for the deaths named by the Harvard group, whose findings supported those of their European colleagues, was a decline in navigational abilities in the bee's brain, in particular the communication of food sources via the tail-wagging dance. This had a devastating effect on the supply situation of the colony and thus on its health and stability.

For years, scientists have held heated discussions about why bees react so sensitively to diverse toxins. A possible explanation goes back several million years; bees have had to struggle against an acquired deficit in the course of evolution. Unlike other insects that feed on green plant matter, bees have not adjusted to the self-defense systems of plants via toxins. Plants can protect themselves from all kinds of predatory herbivores—never perfectly but effectively—through autointoxication. The flowering sections of plants, however, are free from autointoxicants as they do not want to discourage or kill their benefactors, the pollinators. For this reason, in the context of the history of evolution, bees have never had the need to develop a high tolerance to toxins. In the face of today's agrotoxins, this "lapse" makes them particularly vulnerable.

As to the extent of the dangers, one of the world's leading bee experts, Jean M. Bonmartin, got to the heart of the matter in a résumé of his extensive investigations into the commercially available neonicotinoids at Apimondia

2009, an annual meeting of the International Federation of Beekeepers' Associations: The insecticides Imidacloprid and Clothianidin, both produced by Bayer, are seven thousand times more toxic for bees than DDT.

Is that not enough to justify banning the use of neonicotinoids? Not entirely and not everywhere. Two particularly strong sellers, Gaucho and Poncho, are banned in Italy, France, and Germany for use on corn but permitted for use on rape in Germany. In Austria, Clothianidin is also allowed to be used for corn farming, and it is still exported to roughly one hundred other countries. However, Bayer has learned some lessons and improved safeguards. Its own safeguards. In its annual report for 2011/2012, it refrained from publishing the sales figures for Gaucho and Poncho—for reasons related to competition, or so they stated. In the first quarter of 2012, however, non-brand-specific figures indicate that Bayer sold neonicotinoids in the range of 100 million euros (over US$137 million).

Naturally, only the company can answer the question of whether they knew more about the potential dangers of their neonicotinoids than was published. While researching for the documentary *More Than Honey*, Markus Imhoof met scientists who had carried out studies on behalf of Bayer but were not allowed to publish them; a US representative of the German global corporation was "whistled back by one of his superiors as he was giving me information about Imidacloprid, and all of it off-camera," stated the Swiss director, who also had to buy shares in Bayer because otherwise, even with a press card, he would not have had access to the shareholders' meeting. The subject of the controversial preparations was also raised at the meeting. "I stirred up a hornet's nest."

In addition to the complex agrotoxins, other factors are suspected of having an effect on the health of bees. For example, the effects of the ever-increasing number of cell phone towers have long been discussed, although most available studies indicate that the effects of cell phone radiation on living organisms are insignificant. Research findings from 2011 published by the Swiss university ETH Lausanne, however, show that test bees definitely reacted to cell phone waves. When exposed to such waves, they buzz louder and fly away from their hives.

The bee disease researchers' long-term observations show the collapse as a result of a bombardment to the immune systems of these social insects. Or, as Markus Imhoof says in his documentary, "It is the sum of the causes, the bees die as a result of the success of civilization; they die because of us."

Even if we don't share the pessimism of the infamous quote attributed to Albert Einstein—that if bees disappear, the extinction of the human race will not be far behind—the numbers do paint a grim picture. Estimates from the US Department of Agriculture indicate that if the honeybees really did vanish, it would mean annual losses in the USA alone of between US$12 billion and US$17 billion.

For many kinds of fruit and vegetables, there is no pollination alternative to bees. First, only the numerous colonies of *Apis mellifera* are airborne when mass pollination begins in early spring. Their hibernation allows them an energetic advanced start when other pollinators like flies and butterflies are not active. Second, the flower-fidelity of bees is a guarantee that the fruit trees will be full of fruit; the chances are that other pollinating insects will search elsewhere for nectar before a whole tree has been processed. In that case, the pollens of different plants become mixed, which is as much use for pollination as trying to cross a cat with a cow.

"It is the sum of the causes, the bees die as a result of the success of civilization; they die because of us."

"If bees continue disappearing at this rate, it is esti-mated that by 2035 there will be no honey bees left in the US," wrote Alison Benjamin and Brian McCallum in their non-fiction thriller *A World Without Bees*. They also present a possible solution, linked to a warning about seeing this approach as a panacea. Geneticists are currently working on a new, virus-resistant superbee, trying to combine the resistance of some species with the gentleness of others.

Up until now, however, the efforts of the biotech industries have been directed toward realizing the vision of pest-resistant cultivated crops and converting them into hard cash. In 2003, the alarm bells began ringing for beekeepers as Monsanto, the world's largest producer of genetically modified plants, received approval for the use of MON810 in open land cultivation in Germany (repeatedly retracted, as it was at the time of writing, due to environmental risks). The product is genetically modified corn, also called Bt corn, which no pest survives eating. The toxin is embedded in the plant. But, ask the beekeepers, what is the effect on our bees when corn mutates into a potent insect killer?

Such questions are irrelevant, say its producers. They categorically deny that Bt corn is a threat to bees. Professor Hans-Heinrich Kaatz from Jena University published a study on behalf of the Bundesministerium für Bildung und Forschung (the German Federal Ministry of Education and Research) to check the company's claim of harmlessness. He fed a sample group of bees pollen from Bt corn and another group GMO-free foods. The bees on Bt pollen had significantly higher rates of *Nosema* infestation, the widespread, infectious intestinal disease. The other group remained healthy.

Monsanto declared that the Kaatz findings were null and void; *Nosema* doesn't exist in a healthy colony, and if it does, it can be kept in check by antibiotics. "A remarkable statement," says Walter Haefeker, board member of a German beekeepers' association called Deutsche Berufsimkerbund (DBIB) and president of the European Professional Beekeepers Association. "For one thing, *Nosema* is always latently there. What do they mean by a healthy colony? And for another, if all colonies were treated prophylactically with antibiotics against *Nosema*—as we are led to believe from the recommendations of the producer—then even Bt corn would be bee-compatible. It is like saying 'Well, actually our cough medicine does cause a total loss of direction but we also have an effective remedy for that on offer.'"

Apart from the strong suspicion that their bees do not like Bt pollen, beekeepers have another and, for them, far more important reason to make a stand against the Monsantos of the world. Beekeepers cannot sell their honey as pure and free from genetic modification when their bees feed from crops that have been genetically modified. For a long time the biotech lobbies have tried to prevent honey from being subject to the same standards as field crops are. In reality, this is difficult to implement, as outside of greenhouses farming takes place in nature, where winds don't adhere to the borders created by humans, where fields cannot be separated from one another like rooms, and bees are not well known for abiding by the rules of land registry. Farmers have a right for their products to not be contaminated by their neighbors' conventionally sprayed and/or genetically modified field crops.

This right did not apply to beekeepers until the Verein Mellifera founded the Bündnis zum Schutz der Bienen vor Agro-Gentechnik, an alliance for the protection of bees against agro-genetic technology, which financed and organized the claims of a number of beekeepers in the interests of all beekeepers and honey purchasers. After a series of lawsuits lasting several years, in September 2011 the European Court of Justice (ECJ) eventually held that honey with even minute traces of pollen from MON810 corn is contaminated and cannot be sold as a natural product. The alliance continues to fight for the introduction of legislation to ban the use of MON810 corn. So far the simple advice of the authorities is that beekeepers should keep their bees away from MON810 fields while the corn is flowering.

At time of writing, the Bundestag, the German legislative assembly, had not yet turned the ECJ judgment into a valid German law. In Bavaria, however, the minister for the environment, Markus Söder, of the conservative Bavarian CSU party, pronounced that there had to be a three-kilometer (just under two-mile) safety zone around every apiary within which no MON810 is allowed to be cultivated. The density of bees still in existence in Bavaria means that it would be nearly impossible to find areas to cultivate genetically modified corn. This is probably also why Monsanto is trying, by all available means, to prevent coexistence regulations for beekeepers.

Up until now in the USA, the possible effects of genetically modified crops on bees and honey have not been an issue. Large-scale American beekeepers are preoccupied with the problems of their dwindling colonies, a serious threat to the oldest global food industry.

The almond baron Joe MacIlvaine, who year after year hires millions of John Miller's bees, is beginning to think about the unthinkable: What if the honeybees really do vanish? And he doesn't leave it at mere musings; MacIlvaine is privately testing alternatives. Beneath a huge net roofing he is breeding orchard mason bees which to date have proven to be robust against all sorts of rampant diseases. The orchard mason bees are not social insects but solitary. The queen only lays six to ten eggs and hides them in hollow spaces and seals them until they are mature. MacIlvaine is having their breeding biology studied in detail by a beekeeper on his payroll.

Naturally, he is particularly interested in the gathering and pollinating abilities of these stingless mason bees. In the USA they are also known as Indian bees, because they were there long before the settlers introduced European honeybees to the New World; the Native population called these foreign bees "white man's flies."

According to MacIlvaine, breeding orchard mason bees currently costs US$0.10 per insect. This would make it totally uneconomical to breed enough bees for the mass pollination of Californian almonds, especially as the mason bees do not hatch in time for almond blossom and would have to be adapted, via artificial temperature control, to function at a blossoming time that does not correspond to their internal clocks.

John Miller from Gackle does not yet have to fear that the native mason bees will interfere with his bee business. His monopoly is secure for the foreseeable future, unless *Varroa*, CCD, and all the other old and new devils refuse to be driven out. And if doubts do creep up on him every now and then, he has only to think of his great-grandfather, the pioneer, who saw obstacles as challenges to overcome.

BEES IN AN
IDEAL WORLD

T HE DIRECTOR OF *More Than Honey*, Markus Imhoof, had a nickname for Fred Jaggi, who appeared in the documentary: Öhi. This was the name that Johanna Spyri's character Heidi, in the Swiss children's book of the same name, gave to her grandfather. Jaggi was, without a doubt, a natural for the part. You cannot look or speak more like a Bernese Oberlander than this man from Gadmental, in the canton of Bern. In the documentary *More Than Honey* he represents the esthetic, dramaturgic counterbalance to the industrialized, soulless bee world that the movie also features. Jaggi stands for down-to-earth, rural beekeeping with a manageable amount of colonies deep in the unspoiled Swiss countryside.

He worked primarily as a carpenter, and until a serious accident in his early twenties he was a passionate biker. This combination of work experience and hobby had previously brought him into the movie world when he built a hidden ramp in a glacier for a motorbike chase in a James Bond movie. Today he indulges in two hobbies: making cupboards out of tree trunks worn smooth by the frosts of the mountain climate and bleached by UV light, and, of course, bees.

Bees weren't the love of his earlier years—as the child of a rural beekeeper, he avoided them because of their stings. However, his elderly father gave him a choice between joining him and giving them up, and as he says, he didn't want to be a softie. The neighboring beekeepers, as he remembers, "weren't actual beekeepers, they were farmer beekeepers. You looked around at what the others were doing and basically just got on with it as a sideline. Only occasionally did you grab a reference book."

In the 1980s and 1990s, the era of after-work beekeeping was over. By that time the *Varroa* mite was appearing even in secluded alpine valleys, and good honey yields in the mountains could only be achieved with concerted and virtually professional measures against bee diseases; "then there weren't many who carried on."

Jaggi was one of the few who did. The Bernese are known in Switzerland for their stubbornness but also for their devil-may-care attitude to life. The difficulties Jaggi faced brought these characteristics to the fore. And then, as if to comply with all the alpine stereotypes, he started poaching, but only occasionally, you understand, and not necessarily from conviction but rather from a quintessential Swiss stubbornness. There was the time with the badger he shot that he had to hide from the gamekeeper, or with the doe that hung in the smokehouse. Previously he had been refused a hunting permit, partly because of a disability. Not that that stopped him hunting, of course.

He refers lovingly to his bees as "my blacks" or "ur-bees," which fits in well with his fondness for all things primal. Totally different from the others that buzz around, his bees are superbly adapted—to the altitude, the climate, the six-month-long winters, and in particular, to the special diversity of flowers of the mountain meadows. The scientific name of his bees is *Apis mellifera mellifera*, or as beekeepers call them, *nigra*—the blacks.

But even if Jaggi doesn't like to hear it, his *nigra* are in no way originals.

Rather, in the form that they are found in Switzerland today, they are the results of the efforts of the Swiss breeder Ulrich Kramer, who in 1890 tried to "pure-breed" *nigra* in special reserves with the intention of maintaining the original form. Pure-breeding, however, meant strict inbreeding. Breeding trials on honeybees up until that time were almost always aimed at nurturing gathering abilities, gentleness, and swarming inaction. Bees were meant to bring home lots of honey, to allow intrusions into their hives without resistance, and to swarm—when the whole colony goes off in search of a new residency—as seldom as possible.

But as so often happens, the breeder's preoccupation with a few aspects meant that other, equally important attributes were also affected in unexpected ways. Selection through inbreeding has side effects. In the medium term, this breeding practice had an immense impact on the health of the bees, individually and collectively—similar to the practice of breeding cows as living milk tanks, where previously healthy cows, through the greed of breeders, have their immune system destroyed and can only remain on their chronically overloaded legs with the aid of chemicals.

Nevertheless, Jaggi went for racial purity. (In Switzerland, it is more acceptable to use the term "racial purity" than it is in neighboring Germany.) Better they keep themselves to themselves, and here he is in the majority. "You see what happened when they crossed the European and the African bees—suddenly we had killer bees. If something like that happens here, then there will be fun. By then I'll no longer be keeping bees, I'll have met my Maker," says Jaggi, pointing to some spectacular white clouds above the Geisshorn, a mountain in the Bernese Alps. "This will no longer be my problem." Currently, he tackles the smaller problems himself. When he discovered that some of his bee offspring had gray abdominal rings, he concluded that his queen must have mated with *carnica* drones from the neighboring valley. He found the queen among a mass of workers on the comb and ended the "bastard" production by beheading the queen with his fingernail.

In the course of its relationship with humans, *Apis mellifera*, the honey-bearing bee, has split into twenty-five races in different geographical regions. As Jaggi anticipated, the races have adapted to diverse habitats, either alone or through breeding. Numerous subspecies have become established in various European regions: the Spanish bees are termed *Apis mellifera iberica,* the west and central European ones (Jaggi's "blacks") *mellifera*, the Italians *ligustica*, the southeast European ones *carnica*, the Balkan ones *macedonia*, the Greek ones *cecropia*, the Cretan ones *adamii*, the Cypriot ones *cypria*, the Turkish ones *anatoliaca*, the Caucasian ones *caucasica,* and the Morrocan ones *intermissa*. They all bear the species name *Apis mellifera.*

Only three of them are of any significance to international bee breeders: *Apis mellifera mellifera, Apis mellifera carnica,* and *Apis mellifera ligustica.*

Apis mellifera mellifera, also known as the European dark bee or *nigra,* has no problems with damper climates and copes well with the cold, even during long winters—one of the reasons that Jaggi believes them to be good high-altitude bees. After winter they need a little bit longer to reach peak production, a drawback that has made them unpopular among beekeepers interested in performance. *Nigra* bees are also a bit smaller than other strains, but this is not reflected by reduced transport capacities.

> **Bees were meant to bring home lots of honey, to allow intrusions into their hives without resistance, and to swarm—when the whole colony goes off in search of a new residency— as seldom as possible.**

At the beginning of the twentieth century, there was even something of a *nigra* hype among beekeepers, but that died down relatively quickly. One of the leading figures in the Swiss bee world, Hans-Ulrich Thomas, suspected that changes in agricultural usage, for example, land clearance projects, had a favorable effect on the small black bees. *Nigra* particularly thrived at the harvest of buckwheat and the associated field flora (farmers call them weeds). Nowadays, both barely exist, if at all, so *nigra* are buzzing along on the sidelines. But maybe they won't stay there. The strain is low maintenance, which makes them an attractive prospect for many beekeepers today.

Apis mellifera ligustica, the second successful honeybee on our list, have attained worldwide popularity because compared to their direct competitors, they come out of hibernation quickly, achieve a high headcount, and deliver good yields. Beekeepers prize their gentleness, their unwillingness to wield their sting to prevent the theft of their honey, as notoriously executed by humans. *Ligustica* have become established in all regions with a Mediterranean climate, but flag in areas with longer stretches of bad weather. Additionally, experts have discovered that *ligustica* get lost easily, and thus they are a bad selection for the milking of aphids in cluttered forests.

There remains only the third member of this group: *carnica*. They are the most popular strain among beekeepers and worldwide the most widely distributed bees. *Carnica* have inconspicuous gray rings and are slightly bigger than the *nigra*. They originate from the eastern Alps, the eastern Danube area, and the western Balkans, and they have prevailed against the local competition because in every respect, they attain top or nearly top grades for most of the desired qualities. *Carnica* queens hibernate with relatively few winter bees, the advantage being that not so many need to be fed. In spring, the *carnica* winter bees get into gear relatively quickly and can immediately work efficiently. If the regional plant cover or the climate changes, they are able to cope with the new situation much quicker and more effectively than other strains. Also, their lack of aggression has a considerable bearing on their success.

Jaggi inherited the *nigra* from his father, who in turn had inherited them from his father. As an avowed traditionalist, he obviously cannot suddenly consider bad something that until then had been considered good. No, it can't be bad just because others have turned away from them! You don't have to follow every fashion! But now that his "blacks" were being harried on all sides by *carnica*, he was worried about being driven into a corner by the interlopers.

This intermixing of races is possible because of the special reproductive biology of bees. Virgin queens do not mate in the hives but leave the hive and fly in the vicinity of the site where other genes are available. The bearers of these genes are winged—the drones.

But one thing at a time. In a colony there are three kinds of bees: a queen, many male drones, and an even larger number of workers, which are female but are only able to reproduce under very special conditions and then only to a limited extent.

In the course of its five-year life, a mated queen lays millions of eggs in the brood comb which can become drones or workers. A queen that is ready to lay eggs recognizes by the size of the comb cell whether it is the brood chamber of a drone or a worker. Depending on the specifications, it fertilizes the egg using its already filled spermatheca, in which sperm can be stored for years, or it lays unfertilized eggs, the so-called drone brood. Drones develop in these unfertilized eggs with a genetic makeup that is identical to the queen's. If the queen fertilizes the egg it becomes a worker.

But how is a queen reproduced? Like the workers, it comes from a fertilized egg, but one that its mother laid in a very special, differently shaped, and slightly bigger cell. This "queen cell" is prepared by workers as soon as the necessity arises—that is, when the colony needs to swarm. Simultaneously, the old queen is put on a diet so that it loses weight and can fly again, something it has done only once before in its life, for its nuptial flight. Before the new queen hatches, the old queen swarms out with some of the bees to found a new colony, leaving the fully stocked hive to the younger generation and the queen-to-be. Ethnologists recognize similar divisions in social units of primitive peoples. If a tribe becomes too big for the land that is supposed to sustain them, the experienced old tribal chief will move on with half the tribe to seek new resources. A younger chief remains in the old homeland with enough to offer those that remain.

Bee breeders have used this natural principle of bees as a model for a targeted increase in their colonies by "splitting"—dividing a colony into two or more smaller colonies, each one having an implanted new queen or queen larva in a new hive. Splitting on an industrial scale, as practiced by operators like John Miller, is a procedure that causes a lot of stress to bees.

Whether artificially introduced by a beekeeper or reared by the bees themselves, a new growing queen in a hive begins as a normal female larva. The decisive factor in the development of the queen is the larva's diet. In the first larval stage, all bee larvae are fed royal jelly, a mixture secreted from nurse bees' hypopharyngeal glands. While the diet of future workers and drones is soon changed to almost exclusively pollen and honey, the larvae of the future generations of queens receive lifelong supplies of royal jelly. Only in this way can the ovaries develop completely, and only in this way do the larvae become queens capable of reproduction.

When the bee larvae have reached a certain stage of maturity, the nurse bees seal the brood chambers so that the fascinating process of metamorphosis can take place in total isolation, as it does thousands of times each day in the spring and summer in every beehive: A worker, a drone, or a queen is created from a formless entity. The development times vary somewhat. With European bees, the workers hatch twenty-one days after the sealing of the brood cells, the drones after twenty-four days, and the queen after only sixteen days.

Soon after the queen's comb is exposed with the assistance of nurse bees and the new queen hatches, a cloud of sexually mature drones gathers at the nearest assembly point and they mate. This happens mid-flight, and is an intense process with up to twenty partners that pay with their lives for their first and only copulation. The drones' sexual organs are ripped off during coitus. The drones that return to the hive, alive but without having achieved anything, do not fare much better. They cannot feed themselves, which is why in fall they are driven away or killed by the workers which consider them to be unnecessary consumers.

The queen, on the other hand, with a stock of sperm for the fertilization of millions of eggs, has a long working life ahead of it. Between April and September, it will lay up to two thousand eggs a day, for up to five years.

Depending on the site and location of neighboring beekeepers, unwanted genes and thus unwanted attributes can be introduced during the nuptial flight as the male suitors are drones from very different backgrounds. Many beekeepers try to avoid this, and prefer fertilized and pure-bred queens to implant in their colonies, instead of taking risks on the natural marriage market. Hybridity is only one of the dangers that Jaggi is confronted with, and the smallest when measured against the others.

Even in the healthy air and up on the high meadows that have never been and never will be polluted by agrotoxins, bee diseases are rampant. And they can affect anyone, even experienced beekeepers like Fred Jaggi. One day he stood in front of burning frames and hives that he himself had lit. A bee inspector named Elisabeth Schild had discovered foulbrood; at least five of his ten colonies displayed the fatal symptoms. Schild didn't have to explain much. Dead, dark-colored pupae could be seen in the cells and there was an acidic smell. Some nectar forager or other had introduced the deadly bacteria and the nurse bees had spread *Melissococcus plutonius* throughout the hive while cleaning, like a cleaner in a hospital taking a bucketful of germs from one ward to another. In Switzerland, the rule is that when more than half the colonies are affected, all the colonies have to be destroyed to stop the disease from spreading.

For a number of years now beekeepers have differentiated between European and American foulbrood. The European version (EFB), the one that afflicted Jaggi's colonies, is bad; the American foulbrood (AFB) is worse. The name American foulbrood is misleading, as the areas affected are not restricted to the North American continent. The name is simply derived from the chance fact that it was first properly recognized and described in the USA. The disease can be recognized by sunken, capped breeding cells. Typically, well-developed larvae are affected; they change to a sticky slime that is a bit cobwebby when poked with a toothpick. Infested colonies quickly become weak, and members of stronger colonies soon plunder the weakened ones. There is no easier way to procure honey than to steal it—even for bees. However, the raiders carry home not only the bounty but also the plague, like the plundering hordes of the Thirty Years' War.

Officially, within the European Union (EU), AFB cannot be countered with antibiotics. If people persist in doing so, the chances of keeping the virus away from growing broods are minimal. Endospores, inactive and long-lasting forms of the pathogenic bacteria, can survive for decades in the brood comb cells—even as completely dried-up masses on the wooden frames and supports. What is more, endospores travel around the world in honey jars. World trade, to a certain extent, feeds AFB bacteria. At least it is not dangerous to humans—or so the experts tell us.

What is more, endospores travel around the world in honey jars. World trade, to a certain extent, feeds AFB bacteria. At least it is not dangerous to humans— or so the experts tell us.

Unlike AFB, European foulbrood only affects broods that are younger than forty-eight hours. The dead larvae don't become slimy but, as Jaggi experienced, turn into yellowy-brown shriveled objects, and there is an acrid smell. The EFB pathogen dies at 78°C (172°F) which prompts beekeepers who have been affected by infestations, especially in Switzerland, to decontaminate their inventories by baking them at 80°C (176°F) in large ovens. The colonies are beyond rescuing. In the meantime, they have learned other defensive techniques like washing the complete beehive, including the frames, with 5 percent hot caustic soda, followed by a 6 percent hot soda solution, and finally wiping it down with 70 percent alcohol. However, considering the effort required and the fact that there is no guarantee of success, bee experts and agricultural advisors recommend throwing the whole inventory into the nearest incineration plant. Those who live off the beaten track, like Jaggi, are allowed to burn them on their own land in an 80-centimeter (just over 31-inch) deep hole. Prior to this, the bees are gassed in their hives and the business site is declared a restricted area.

"So there you stand, the burning honey is sizzling, and the dead bees crackling in the flames. It was the same for the farmers who were hit by foot-and-mouth disease and had to get rid of their animals. I've gassed ten colonies and destroyed over three hundred frames. That's three thousand francs [around US$3,000] up in smoke. There was fire in the hole for three days. It was hell," Jaggi remembers. Nevertheless, he didn't give up. After the ban had expired, he fetched three new colonies of bees from Brienz on Lake Thun, an area without foulbrood.

You don't want to experience something like that twice. Despite a successful restart, the question remains: "What did I do wrong? Did I do anything wrong at all?" The official line was that a strong colony with enough "specialists" taking care of hygiene, cleaning bees, was pretty safe. Did some of Jaggi's colonies come out of hibernation weakened? Had he possibly used too much formic acid against the *Varroa* mites and thus indirectly weakened the growing larvae? Jaggi complained that foulbrood research was going nowhere. The scourge strikes, everyone ducks, counts and burns the dead, and hopes that the next time they are spared.

On a morning late in June, clear and warm despite the altitude, Jaggi watches his black bees, the new ones from Brienz. They fly to the alpine roses like red garlands festooning the slopes. Jaggi listens to the buzzing sounds. Even after all these years he still cannot get enough of observing them flying from blossom to blossom.

The watery liquid secreted by flowers, or more precisely from special glands in the calyxes, is initially just a raw material. Nectar is as much honey as flour is cake. But nectar is a highly remarkable substance with a very special composition. Apart from its own minerals, it contains various kinds of sugar including fructose, glucose, and sucrose. Depending on the type of flower, the sugar concentrations can be anywhere from 25 percent to 75 percent.

The plants use two tricks to help consumers like bees and bumblebees find the coveted foodstuff; during the blooming period they lure the insects with fragrances, which are also a component of nectar, and with attractive colors. Without these two stimuli, nature's approximately 130-million-year-old practice of giving insects a key role in the sexuality of plants would not work. Insects, and above all bees, cannot reach the energy-rich juices of the blossom without brushing their hairy abdomens against the pollen on the stamen, the male reproductive organs of plants. The pollen dusts the small flying objects so that on the next visit to the same kind of blossom, the stigma, the female reproductive organ, is brushed. The result is pollination, without which apple blossom wouldn't produce apples and raspberry blossom wouldn't produce raspberries.

Wind pollination is an alternative to the interaction of plants and insects for which the flowers need neither brilliant colors nor pleasant scents to attract attention. It is, however, much less efficient as it relies on the mass production of pollen because only a fraction of the pollen that is spread throughout the landscape lands on a matching flower. In contrast, insect pollination is a process of utmost precision and is pretty much independent of weather conditions. Only nonstop rain or a cold snap at an inopportune moment could temporarily bring the ingenious system to a standstill.

Pollen is by no means tiresome ballast for bees but rather an important foodstuff. Nectar alone is not enough to sustain the brood and colony throughout the winter. Pollen contains fat, proteins, and minerals as well as vitamins. Without fattening up with this diet the bee larvae cannot develop their organs and glands. The worker bees use pollen to build up their strength for the hard work ahead of them. Pollen is also very important for the rearing of drones. Only well-fed drones are strong enough to win the battle for mating with the queen.

Before flying back to the hive after foraging and pollinating, they systematically brush the pollen off parts of their bodies and then redistribute it. It is important for pollination that this activity happens after the bees have already offloaded some of the pollen onto the female stigma. During the cleaning activities, the bees brush pollen that has settled on their large, finely haired compound eyes and frontal thorax toward the abdomen with a special little brush on their forelegs. The middle legs then take on the task of passing the pollen toward the specially adapted hind legs. A highly functional arrangement of special hairs, pollen combs, and pollen presses work together until the bulk of the pollen has been compacted into pollen pellets and attached to pollen baskets on the tibia of the hind legs. A full pollen basket can weigh up to eight milligrams and a twin pack (that is, a basket on each side) sixteen milligrams (each of which is a tiny fraction of an ounce). The antenna cleaner is an especially ingenious part of the bees' cleaning and transport features: semicircular notches in the forelegs through which the worker bee's antenna is drawn in order to remove pollen and debris.

To extract the nectar fluids, the bees are equipped with a proboscis, or rather they have one when they need one. They adjust their mandibles and labial palps (roughly corresponding to our lips), making the hairy tongue into a tube at the end of which is a sort of spoon, giving their tongue the form of a nozzle. It then moves back and forth, functioning rather like a suction pump, so that the fluids are lapped up and are fed, via a fairly long esophagus, into the honey crop. The common term "honey stomach" is misleading as nothing is digested here, only stored. The nectar can then be unloaded back at the hive. A pyloric valve separates the honey crop from the actual digestive tract in the abdomen, where only small amounts arrive to be used as fuel for the flight.

On top of this—and enormously important for the survival of the colony—the bees can use this valve to shift certain pathogens and toxins from the honey crop to the ends of their own bodies.

When the bees return to the hive they have no time to rest; they immediately begin to offload their supplies. This is hard physical work. The returning forager regurgitates the contents of the honey crop, passing it on to a home-based worker which in turn passes the nectar on to another colleague. This relayed transport means that by the time the nectar reaches the comb for storage, it is more concentrated, more fermented, and thicker. The returning foragers stow away the pollen themselves without bothering the intermediate workers.

The diet of the worker and drone broods, after their initial diet of royal jelly, is enriched by food from the pollen, nectar, and honeycombs after their fourth day of life. Any nectar that the nurse bees don't immediately consume is made suitable for storage by the other hive bees by sealing it in the combs. Once sealed, the nectar becomes honey as we know it.

Fred Jaggi's honey is not just any old honey. "Up here the bees don't have the option of going for sprayed crops. No artificial fertilizers near or far. My honey is organic, more than organic—absolutely pure! The alpine roses are extremely good. When they flower and there's good flying weather for ten days then our harvest is assured. Alpine roses produce this wonderful, almost crystal clear, very runny honey. Anyone who tries it stays a customer. I don't need to bother advertising."

Those who wish to relieve the bees of their laboriously gathered goods have to invest a bit of effort themselves. The box-shaped dwellings of the bees have to be opened and the frames with the honeycombs removed—and the colony does not give up its honey without resistance. But Jaggi has figured out something that calms them down slightly. "The evening before harvesting, I gather some valerian roots, chop them finely, and place some of them near the entrance holes. The returning bees carry the scent into the hive and it spreads through the colony. The bees are not drugged then but when I collect the honey they are docile." Well, others would say less wild. "I have to put up with ten, maybe fifteen stings per day from my bees without grumbling. It's like being scratched when picking blackberries. But there are also things that force me to stop immediately, for example, nearby helicopters; the bees become aggressive and there's nothing you can do about it."

That they don't react too aggressively to the removal of the honeycombs can be put down to the century-old practice of selective breeding by breeders who valued gentleness. But even these colonies are not boundlessly peace-loving, which is why beekeepers use defensive fumes puffed out of smokers, as well as protective clothing, from gloves to hats with netting to full suits.

While for numerous beekeepers the honey yields are the focus, all too often at the expense of the bees, beekeepers working with the anthroposophi-cal Demeter Association have shown that profitable harvesting and a belief in the welfare of the bees do not have to be contradictions. Their self-imposed principles of bee breeding, which also take the natural habits of the insects into consideration, differ, in some respects fundamentally, from those of tradi-tional beekeepers. Their core philosophy in a nutshell is: Let the bees do what they like doing as often as possible and refrain from doing what they don't like as often as possible.

Before mobile units with moveable frames were introduced in the nine-teenth century, the bees kept themselves to themselves in their tree trunks, their clay tubes, or beehives until the moment when someone other than themselves came after their honey. The beekeepers had to break open the hives and thus destroy them in order to gather the harvest. Only since the advent of the box system, which permitted the removal of individual frames, have beekeepers been able to follow procedures within the beehive—and to interfere with them. Conventional beekeepers have the bee colonies live in "prefabs" made of industrially produced wax with defined cell sizes for worker broods and honey cells in which to implant the honey. The architecture of the hive is predetermined. Demeter bees, by contrast, can construct their combs much as their collective sense chooses. When frames are inserted, they func-tion as units into which the bees can incorporate their own natural combs.

The principle of beekeeping according to the nature of bees, which Demeter beekeepers follow, allows the bees more opportunities for development and variation. A queen is not imposed on them, they are allowed to swarm and choose a natural form of reproduction. The colony decides when to split and form a new colony. Demeter beekeepers are strictly forbidden from clipping the wings of queens, a practice that facilitates the capture of a swarm, as the disabled queen falls to the ground in front of the hive and the swarm remains there with it. And like humans building eco-friendly homes, the residents of the hives should only come into contact with untreated woods, not with synthetic materials like polystyrene. The highly dangerous *Varroa* mites are solely combated by non-synthetic measures like lactic acid or oxalic acid.

In essence, the principle is: Eat healthily. Even Demeter beekeepers cannot guarantee that the surroundings where the bees fly are unsprayed—their foraging radius is more than three kilometers (around two miles). Preferred sites are, however, those bordering on areas that are organically farmed. Demeter bees are also usually allowed to keep more of their honey than are colonies of conventional beekeepers; 10 percent of their own honey must remain for their winter stocks. Behind all this is the conviction that their own product strengthens the animals more than sugar water and the industrially produced bee foods that are usually fed to bees as substitutes for their own plundered stocks. According to the Demeter beekeeper Günter Friedmann from Küpfendorf-Steinheim, southeast of Schwäbisch Gmünd, Germany, this form of beekeeping is so successful that more and more bee breeders are prepared to change to this approach. "We have an influx, and this wouldn't be the case if we were just being bee-friendly, but couldn't show profits."

Only since the advent of the box system, which permitted the removal of individual frames, have beekeepers been able to follow procedures within the beehive—and to interfere with them.

In 1923, Rudolph Steiner, the founder of anthroposophy and so a mentor for the Demeter beekeepers, stated that in eighty to a hundred years—that is, today—beekeeping would be confronted with a great crisis, especially as the bees were being denied their natural reproduction by the implantation of an artificially bred queen. "Certain forces which have hitherto been *organic* in the hive will be mechanized..." According to Steiner, "we have to be aware that by working mechanically we destroy what nature has elaborated in such a wonderful way."[1] The words of the esoteric Steiner are not based on knowledge about "inbreeding depression" or "genetic impoverishment," instead they sound more like a general warning not to violate nature and its creatures.

In comparison to the drive for maximum efficiency in beekeeping and breeding, as is mostly practiced nowadays, the veneration shown to bees by long-gone cultures over millennia now appears rather strange.

Bees were working animals of special significance for humans from the earliest of times. A bone carving dating from around 4000 BC was discovered in Bilche Zolote in Ukraine. It shows a stylized bee goddess on a bull's head, an early link between cows and bees that was to be found centuries later on ancient Greek gems and coins, and centuries later again on wooden carvings in the Middle Ages. The Swiss bee researcher Matthias Lehnherr interprets this striking combination as a kind of plea for prosperity. "Cows, bees and beehives symbolize the Promised Land, the land of milk and honey."[2]

In ancient Egypt, beekeeping and even migratory beekeeping was known around 3000 BC. There are portrayals of honey raiders, and of beehives being sailed up the Nile, presumably following the blossom. As honey was for many millennia the only source of a sweetener for humans, it might also explain why the Egyptians held the small working creatures in such esteem. They were valued so highly that they were accredited with the highest possible origins. According to ancient Egyptian myths, bees were created from the tears of the sun god, Ra. In ancient Greece, bees represented divinity, vitality, and wisdom. The freshly born Zeus and Dionysus were nourished with honey. Dionysus, a god better known for excessive wine consumption, was even supposed to have taught humans the rudiments of beekeeping.

And even in the Bible and in the allegories of the Church fathers the bee has a special status. Flying heavenward, they symbolize Mary, Queen of Heaven, who conceived the Savior as a virgin, rather like the bees that are born "from wax," apparently asexually. The thought of swarms of drones having group sex with a rather unchaste queen somewhat tarnishes the traditional story of Mary's virginity. Bees represent reincarnation. Just like Jesus, who remained shrouded in mystery for three days after his death on the cross, the bees are invisible for the winter months only to rise again fresh as ever in spring. In the Middle Ages, bees symbolized the two sides of the son of God: sweet as honey and mild is the mercy of the Lord; painful as a bee's sting is the verdict of the Judge of the World at the day of reckoning.

The association with holiness remains in particular with wax, an integral part of many liturgical acts, from the thin intercession candles to the stately light columns at Easter. Even today candles are an indispensable part of ritual acts for a variety of religions. Of course, today most of them are produced industrially and are made of stearin.

The cultic veneration of bees survives to this day in popular sayings. As they say in Switzerland, "Curse and swear where a beehive stands, the bees'll sting you on your hands."

BEES IN
THE LAB

THE HONEYBEE IS the only one of around a thousand kinds of bees in Europe that has developed an extraordinary form of a differentiated society, with two of its subsystems being storage management and preparation for winter. Unlike, for example, wasps (*Vespinae*) whereby only the well-hidden fertilized queen survives the cold winter and in spring has to build up its retinue from scratch, the honeybees do not depend on a single individual for survival. The strategically reduced population of *Apis mellifera* (the colony is reduced to preserve food stocks) makes it through the winter, and so they are ready in early spring, when it counts, to pollinate our early blossoming fruit species.

With their ability to start work in large numbers in the cool early days of spring, the honeybees have a special status. They fly, if they have to, at outdoor temperatures of 8°C to 10°C (46°F–50°F), though their optimum is between 20°C and 25°C (68°F–77°F). Bumblebees can even fly at temperatures of around freezing point, which gives them a considerable head start in mountainous areas.

Later in the year, bees are only one of many insect groups and at that stage they have to share pollination duties with beetles, butterflies, flies, and a number of other specialists. Although bees are pollinating generalists, meaning that they are not reliant on a few flower species or forms, they are "flower-faithful." If, for example, a bee discovers a blossoming apple tree and has begun to gather the pollen, it will remain there and harvest the resources. It will only change its food source when the present source no longer has enough to offer, irrespective of how tempting or fragrant the other varieties of blossom are. They know the time has arrived when, among other things, it is difficult to unload the nectar that they have gathered on the receiver bees at the hive entrance. The flower-fidelity of the bees is a crucial prerequisite for monofloral honeys to be produced outside of pure monocultures. Bumblebees also learn the respective varieties of blossom, but forage with a strategy that still leads to diversification: Every one of them regularly checks whether it is foraging in the best possible way by occasionally flying to other kinds of blossom.

Each insect has its own particular area of responsibility and deals with very specific plants. Botanists talk of plants that exhibit cantharophily (that is, they are pollinated by beetles), many umbel blossoms, elderberries, privets and herbal shrubs; myophily (pollinated by flies), such as black or white veratrum; and melittophily (pollinated by bees), like most labiates and rosaceae, to which apples and pears belong, as well as blackberries, cherries, plums, and almonds. This division of flora would not continue if the insects failed to recognize the blossoms of their choice and just flew off on haphazard foraging/pollination tours.

Insects can see colors within a certain light spectrum, but they are different from those that humans perceive. So, just as it is difficult for colorblind people to convey what they see to people not suffering from color blindness, we cannot envisage how bees and other insects see color. Our retinas have blue-, green-, and red-sensitive cells. Bees have green, blue, and ultraviolet receptors; red for them is only something dark. Because they can see ultraviolet, bees can see lines, color patterns, and mixtures of colors on some blossoms that we cannot discern. A blossom that looks pure white to us is only white to them if the UV light components are included and the red components are missing; they see yellow blossoms with UV reflection as a bold "bee purple." Their perception of colors is not less pronounced than ours, just different. And our honeybees do react to color signals but, as a rule, do not allow themselves to be lured by the most colorful options.

Strengths like getting off to a good start in the blossom season, which is possible thanks to collective hibernation, their flower-fidelity, and their pronounced color vision would not alone be enough to make bees the proverbially busy gatherers that we know and love. Bees learn particularly quickly and effectively, and optimize their work on the basis of what they have learned. A significant prerequisite for this is their exceptional capacity for navigation.

In order to understand this, Professor Randolf Menzel from Freie Universität, Berlin, has been studying the wings and brains of bees for thirty years. The seventy-two-year-old knows his bees like no one else, and he loves them. Menzel can relate to the godfather of all bee researchers, Karl von Frisch (1886–1982), who is reported to have said that a bee colony is like a magic well—once you have drawn from it and the surface resettles you realize that it is bottomless. This is exactly what Menzel has experienced during his decades of research.

The nature of science lies in dividing marvels into their individual components: How do bees fly? How do they reach their destinations? Which of the very diverse signals do they need to receive and process to navigate successfully?

The flight of honeybees is humdrum and yet fascinating. The bees of a colony, adding together all the individual forays, travel roughly three times around the world for one kilogram (just over two pounds) of honey. One single bee collects a teaspoonful of the coveted sweetener during its stint as gatherer. Cruising speeds of twenty-five kilometers per hour (15 miles per hour), with top speeds even reaching fifty kilometers per hour (thirty-one miles per hour), and up to 280 wing beats per second have been recorded.

How is this possible? The thorax of bees mostly consists of flight muscles. But this alone cannot explain how in a straight flight they can cover a hundred-meter (328-feet) stretch quicker than a world record–holding human sprinter. The comparison with another competitive sport clearly demonstrates this: Even the strongest rowers, irrespective of how much traction per stroke they can generate, could not achieve comparable motions and certainly not the necessary frequency. The power transmission according to the laws of leverage is just not enough.

The propulsion of hymenoptera (bees and wasps), flies, and mosquitoes functions very differently. They do not need two muscles, one pulling up and one down, for every beat of their wings. They use their muscular strength to create a chitin soundboard, the entire middle section of their body vibrating rather like a guitar string. The oscillations are transferred to the wings by an ingenious coupling joint that translates them into movements that are quicker than the human eye can register.

The flight of honeybees is humdrum and yet fascinating. The bees of a colony, adding together all the individual forays, travel roughly three times around the world for one kilogram of honey.

This ability has caught the attention of bionic researchers, a cross between technicians and scientists, investigating which of nature's inventions are technically adaptable. How can something so small perform to this level with astonishingly little evidence of wear and tear? How exactly does the "patent" of hooks catching a ridge on the lower margins of the front wing allowing the bee to link the fore and rear wings work? The relationship between energy and performance in this efficient use of muscular power is also remarkable. Bees, however, have to refuel intelligently before their forays; they should start with enough energy and carry as little weight as possible so that they can fill their honey stomachs. Additionally, the possibility of an unsuccessful flight has to be taken into account, so they need enough energy to fly back to base.

Possibly even more fascinating than the pure flight mechanics and energy efficiency are the navigational abilities of bees, the question of how bees find their way around, in particular how they find their way to one or more of their food sources, and above all, how they eventually find the way back to the colony safely.

A tiny brain consisting of only about 1 million cells receives, transmits, and switches from one information system to another. How does that work? Does all this function according to a preprogramed automatism? Are they just stimulus-response models, established processes triggered by an appropriate stimulus? No, says Professor Menzel, a rigid program is not possible; bees have to be flexible to a certain extent.

Menzel has systematically researched areas where there are both scope and options that make the bees "competent" to act—particularly their repeated visits to blossoms. He did not have to start from scratch for his studies on the navigation and learning capacities of bees. As early as 1910, Karl von Frisch had demonstrated that bees could learn to recognize a color or a scent if they were rewarded with sugar solution for associations. In the 1950s, one of von Frisch's assistants transferred this food training to individual bees that were kept in tubes so that their behavior in response to a stimulus could be precisely observed—a kind of Pavlov's test for bees. (A reminder: The Russian physiologist Ivan Pavlov conducted an empirical study at the beginning of the twentieth century in which the feeding of his dog was accompanied by the ringing of a bell. After a while the animal reacted to the ringing of a bell by salivating even when food could not be seen or smelled. The phenomenon of classical conditioning had been discovered. A seemingly neutral signal becomes a key stimulus when it is regularly linked to certain conditions.)

Menzel also used these methods for his studies of bees, which he placed in small tubes and stimulated alternately with the scents of geraniums and roses. One of the two scents was rewarded with sugar solution, which led to an extended proboscis for this scent, even when subsequently there was no longer a reward. Menzel opened the head plate of the bees and took photographs with a special camera so that he and his team could monitor and localize the processes in the bee's brain during the cognition and learning of the conditioning of a scent. Using florescent dyes, they were able to display which brain cells were activated by scents and in which area of the brain this was taking place: red signaled particularly high activity, and blue inactive nerve cells. Having been fed exclusively in association with the scent of geraniums, the test bees only reacted to these smells. Presented with the scent of roses, certainly still appetizing but which had proven to be disappointing, they remained passive.

From the Pavlovian bee test it can be deduced that bees are not solely pre-programmed creatures. Their reflexes can adapt to individual experiences.

Those wanting to see Menzel at work outside his lab have to travel in summer to Klein Lüben in Brandenburg where he has set up a kind of open-air test track for bees. The landscape is not exactly what you would expect to find in tourist brochures; it's a little bleak and short on distinguishing features. However, this is just what the scientist wants and it suits his purposes perfectly. He wants to make and remove the features himself to measure the effect on his flying test objects. It would have been more difficult to say what the individual bees were reacting to had there been other competing landscape features. On top of this, Menzel chooses a time when there are few other blossoms on offer so that his flyers can totally concentrate on his test feeding points. The open landscape, which allows for trouble-free trial conditions, was not the only piece of luck for the bee researcher—telemetric technology has made significant progress in recent years and his experiments also profit from these advances. In the 1960s, scientists started attaching bulky transmitters to larger animals, such as red deer or bears, using the signal to track and record their migration routes. In this way, wildlife biologists using even more cumbersome antennae discovered how red deer use their territory both daily and throughout the year: where they eat, where they sleep, or how quickly they settle down after being disturbed. The first birds that were strong enough to carry the smaller version of the transmitter in flight were wood grouse, cranes, storks, and swans. Thanks to these pioneers of telemetry, ornithologists were able to gain new insights into flight paths, winter habitats, and traveling speeds.

Recently it has even become possible to fit insects with antennae so that they can be tracked by a special radar unit. Menzel's bee antennae weigh twenty milligrams (a fraction of an ounce), and at twelve millimeters (a fraction of an inch) are considerably taller than the bee itself. Although a bee only weighs between eighty and hundred milligrams, the weight of the thin wires is, according to Menzel, relatively unproblematic—bees carry up to sixty milligrams of nectar and pollen to their hives per foray. The weather conditions are critical, though: the test bees cannot fly at wind speeds beyond thirty to thirty-five kilometers per hour (18–22 miles per hour). Two to three hours' flying time and a total of up to forty kilometers (twenty-five miles) is standard for the test bees, with the altitude varying between one and ten meters (three to thirty-two feet) above ground level.

When Menzel began his experiments on bee navigation, he already knew a considerable amount about its hows and whys. Bees are flying data gathers and processors. It was recognized that bees had special senses at their disposal, and that there was a certain hierarchy of the sensory impressions: Optical information is the key; route markers, color signals, and structures are pieced together, making and updating a "map" in the bee's brain where they can be accessed when required. They create, as it were, a matrix into which other information can be slotted.

An important tool in this process is the sun as compass. Although the point of orientation on which this compass is aligned changes its position during the day, the bees can recognize directions. And they can use this complex navigational instrument even when it is almost totally overcast. A small window of clear blue sky, no matter where, is all they need to determine the position of the invisible sun. Depending on the height our central star has just reached, typical patterns of diffused light appear in the upper atmosphere, appearing in the sky like wallpaper moving with the sun which can only be seen by those with ultraviolet vision.

The bee's eyes can do a great deal more to make life at breakneck speeds possible. They enable bees to estimate distances from the movement patterns that appear on the structure of the ground beneath them. On top of this, insects must quickly recognize movement, as it could be coming from a rapidly approaching predator. Their compound eyes can do this much more effectively than camera-type eyes, which humans have, can. Whereas humans can only differentiate a few individual images per second, bees can register 265 images. For our eyes, a walking pace of sixteen images per second just gives us the perception of general motion. A bee would see at more than two hundred images per second a series of jerky individual images. This is why they identify a quick and hectic defense as movement and thus a threat. This is also why it is better and safer to repel them with slow tai chi–like movements than with rapid swatting. Dragonflies, the fastest and most acrobatic fliers of the insect world, process motion information of well over three hundred individual images per second and are even more visually refined than bees or wasps.

The compound eyes of insects, enabling almost all-round visibility, consist of many separate single lenses—ommatidia—on a bulging spherical or oval surface. A single lens in the lower area of the eye's curvature can recognize a rapidly approaching flying object coming from below. A split second later, the flying object enters the field of vision of another ommatidium further up on the curvature. The time lag between the two warnings arriving at the insect's brain is used to calculate the speed of the approaching object—similar to the operating principles of the old-fashioned speed traps where the speed of a vehicle can be calculated by using the time taken between two fixed points.

Much has been written about the compound eyes of insects, but little about the olfactory organ of bees. Using their "noses"—tiny pores on their feelers covered by delicate membranes—bees can, unlike us, accurately smell in stereo. They recognize and distinguish scents and the direction from which they are coming—a particularly useful aptitude for close orientation, especially as the insects' eyes, being receptive to motion information, can barely recognize the shapes of flowers and blossoms from a distance of one meter (three feet). Visual flight for bees is also always a "nasal" flight.

On top of their highly sensitive optical and olfactory senses, bees have a couple of other special abilities. A sense of electric fields enables them to reliably reach their destinations. The Johnson's organ on their antennae—a similar structural principle to that of air pressure gauges in aircraft—measures the oncoming air and conveys to the brain when and how flight attitude or the number of wing beats need to be adjusted.

Bees can also detect the Earth's magnetic field, although exactly how they do this remains unclear. Two theories are that it is down to magnetic crystals in their abdomens or that it is simply another special feature of their compound eyes. At the moment we know very little about what the honeybees use their sense of magnetism for. The orientation of the honeycombs is clearly determined by magnetism and the waggle dance is also thought to be influenced by magnetic fields. Whether and how it is incorporated in navigation remains unanswered.

When a bee's brain is functioning without any disturbances, the navigational systems can be changed according to the situation, rather like the captain of a ship putting aside his telescope in sea fog and switching on the radar. Having this option is crucial; if bees were only out and about in optimal, clear flying conditions they would not be the best pollinators in the world—and we know that they are.

The most remarkable form of communicating information—the famous waggle dance—had already been observed by 1828 by a German beekeeper named Nikolaus Unhoch: "It may seem ridiculous to some, maybe even unbelievable, when I claim that bees also [...] enjoy a certain revelry and pleasure amongst themselves, so much so that according to their nature they engage in a special dance. [...] What this dance actually means, I cannot yet explain, whether it is a bold gleefulness or encouragement [...] will have to be settled in the future."[1]

Unearthing the meaning of the dance took almost a century. At the beginning of the 1920s, Nobel Prize winner Karl von Frisch meticulously researched and described the significance of the dance. He discovered that the returning nectar foragers that had found a good source of food broadcast it by doing the characteristic circular dance, which could be translated as "Look! I've got some good news!" They interrupt their rondo every now and then to share out tasters from their source. Once interest has been aroused, they disclose, by means of a precisely choreographed waggle dance, where the nectar can be found.

By wiggling its abdomen, a bee communicates pointers about the direction, distance, and quality of the food source; the directional information of the dancing bees refers to the angle between the sun and the food source. If, for example, the objective is a tempting cherry tree in full blossom east of the hive, the bee's tail will point vertically when the sun is in the east. About three hours later, when the sun is in the southeast and the cherry tree is to the left of the sun, the tail position would then alter to forty-five degrees. The distance is communicated by the duration of waggle runs that are carried out in semicircles.[2] The more vigorous the waggle the more worthwhile the food source. The amplitude of the acoustic and sensory signals during the waggle dance might be passing on further pieces of information that go beyond the parameters of distance, direction, and delectability, but we can't be sure. There is much that we do not know yet about the many details that are hidden in the dance information and which are probably directed to sensory channels that we are not yet aware of. We are certain, however, that inside the hive an important role is played by a kind of hearing with feelers. Bees in the vicinity follow the movements and sounds of the waggle dancer very closely and can then fly off to the specified destination. Should they also find the source rewarding, on returning to the hive they dance the corresponding dance. So, a bee subsequently turns something that was passed in complete darkness into directional accuracy in the open air and light. It is able to translate directional indicators that were passed on to it vertically (in the perpendicular honeycombs) into a horizontal landscape outside. That requires a simple form of abstraction abilities.

In order to learn more about the famous sun compass of the bees, Randolf Menzel and John Cheeseman, from the University of Auckland, did something that all scientists like to do when they want to understand how something usually works: they interfered with the usual process. Using isoflurane, they anesthetized bees that were returning to the hive from a good source of nectar. On waking up and trying to fly back to that food source, the bees began by flying off in the wrong direction—but according to their former information it was the right direction. Something in their brains "knows" that the position of the sun shifts clockwise some fifteen degrees per hour. The test bees behaved correspondingly, but as if there had not been any movement in the sun during their anesthetized phase. The isoflurane had stopped their internal clocks.

"Bees can memorize navigation, that much is clear," says Menzel, smiling, "but that is only a rough description. Is it only—and I deliberately say 'only'—a route that they themselves have in their heads? When they fly back to their hives from a fully laden cherry tree are they just rewinding the tape of the route they have already flown?"

An experiment proved particularly revealing in the search for answers: Menzel and his team set it up so that a marked bee fitted with a transmitter—let's call this bee Red 23—was witness to a waggle dance. This meant that Red 23 was in the immediate proximity of a sister-bee that was communicating information about the distance and direction of a particularly good food source in a bee-like manner. Red 23 had previously foraged from an excellent source which was in the meantime running dry. This was why it had returned to the hive and observed the dance of other bees.

Where would Red 23 fly to next? Could it trust the information passed on by the dance, even give it precedence over its own knowledge? Red 23 "decided" to return to its own tried and tested food source. In the meantime, Menzel's students had cleared away the food source without a trace. A little later, the team eagerly followed Red 23's transmitted signals on a monitor. Clearly irritated and indecisive, the bee circled for a while above the site which, on the basis of its earlier memories of a successful foraging flight, it had quickly and correctly refound and which again had proven disappointing.

It was now expected that Red 23 would either fly back to the hive to gather new instructions or set off independently on a search for another, completely new food source. It did neither one nor the other. It flew directly to the site pointed out by the waggle dance shortly before its departure. The information received from the waggle dance that it had not used was not only available later but would let it take the direct route to the new site although it had only been described to Red 23 in relation to the hive.

This meant that it used information about the terrain itself and not from the flight there. What makes this remarkable, to put it in human terms, is that it either must have carried some form of "calculation" or could access a memory somehow resembling an inner map of the landscape. The "calculation" could only have been made if the bee took into account both the paths—from the hive to the food source and from the hive to the site indicated by the waggle dance—relating them in such a way as to determine the direct path between the two sites. In mathematical terms, this is vector integration. Such calculations are comparable to trigonometric operations used in surveying.

But maybe bees have a completely different procedure at their disposal? Namely, a concept of landscape in the form of an inner map. Using this, bees could successfully travel by the most direct route from any site in their memories to another.

By wiggling its abdomen, a bee communicates pointers about the direction, distance, and quality of the food source; the directional information of the dancing bees refers to the angle between the sun and the food source.

What is the upshot of the experiment with Red 23? Red 23 behaved in a way that leaves an observant person with little choice other than to assume some kind of decision-making process: "No, I'm not going to go to the source that that bee's waggling about; I'm off to my tried and trusted site." And immediately afterward, Red 23 "revisited" that "decision," performing an unarguably unusual navigational feat by flying to a site as if it were using an inner map. How this is structured remains a puzzle.

All things considered, it is legitimate to ask whether creatures with brains the size of pinheads "plan" their actions with the aid of map-like memories. But what could be meant by "planning" for insects?

That insects have a certain kind of scope for decision making or a spatial power of imagination stretches the bounds of human imagination. Menzel stiffens a bit, and now has to choose his words carefully. Terms like "imagination" could encourage the research competition to criticize him for trying to "humanize" bees, in Menzel's field a damaging slur on reputation of the highest order. "What we do know and can say is that bees have a memory, that they use this memory to navigate and by doing so can factor in their own experiences and information that they have gathered from their base. And that they can make decisions."

Now that he has uttered the magic words, Menzel clarifies: "When we say that they make decisions between various options, we mean natural processes in bees' brains which have nothing to do with what we would call human, considered knowledge. Cognitive psychologists speak of implicit knowledge, knowledge that is automatically at our disposal." But knowledge can take many forms, and we can say that bees have moments of decision, certain degrees of latitude—will one or the other flight plan be adopted or even none of the already known routes? If one such decision later proves to be unproductive, earlier knowledge that had not been previously used can still be deployed, it is still available. It is not knowledge as we define it in everyday life but what Menzel calls a kind of "bee knowledge," to differentiate it from other forms of knowledge.

If individual bees have such latitude in making decisions then it raises the question of individuality in the swarm. Menzel thinks for a long time. He does not want to make any defining statements about it, but he has observed in the course of his food training experiments that not all the test subjects behave in the same way, that within the social fabric of a colony in which conformity plays a pivotal role there is no such thing as total conformity. He points at an array of scent tubes: "Look at this! When I've trained thirty or forty bees then I know that that one scrabbling around is Blue 59; it always moves in a slightly different way to the others."

US colleagues of the Berlin bee neurologist express themselves more emphatically: "Our results say that novelty-seeking in humans and other vertebrates has parallels in an insect,"[3] said Gene Robinson, an entomology professor from the University of Illinois.

Robinson's team concluded that scouting bees, which scout out new sites for the colony, are especially brave or adventurous. As soon as a swarm has set-tled somewhere temporarily after having left its original quarters, scouting bees survey the vicinity, usually in bands of around a hundred individuals. The scouting bees also prove to be more active and successful than the majority of their sister-bees in the day-to-day bee activities, like discovering and commu-nicating food sources.

The research team compared the brain activities of the "brave" scouting bees and the comparatively unremarkable remaining bees. "The magnitude of the differences was surprising given that both scouts and non-scouts are for-agers," the Illinois team summarized.

Recruits in the field, and experienced foragers whose hive has been moved, explore the surroundings in an orientation flight, not yet interested in for-aging for nectar and pollen. They make large loops around the hive, creating a kind of memory of the landscape, slowly increasing the scope, first nearby and then further afield. These exploratory flights are essential for survival. The bees' accuracy in finding the "right" blossoms is not a crucial factor, but it is essential that they know the environs and can find their way back to the hive. A mistaken flight to an alien colony could prove lethal—the guard bees there would make short work of any foreigners emitting the wrong signals. In addition to making orientation loops, before leaving a food source the foragers make a reconnaissance of its surroundings. They memorize colors and shapes, thus making their next visit easier.

Martin Giurfa and Aurore Avarguès-Weber from Toulouse University are also interested in the apparent contradiction that primitive insect brains could have learning capacities. They put forward a working hypothesis that bees can not only recognize and remember a pattern but that they can also find it even if, to a certain extent, it deviates from what they have learned. Using sugar solution as a reward, they trained bees not only to differentiate simple dot-and-dash face-like representations from other patterns of dots and dashes but also to prefer them. After a short time, the test bees flew confidently and reliably to the sketched faces. This also worked when the faces were hidden in complicated patterns and even when they were embedded in enlarged passport photos. However, bees, unlike humans and many other higher animals, cannot distinguish individual faces. The dot-and-dash diagrams that the bees recognized could possibly be the equivalent of an idiosyncratic kind of blossom that they have recognized and memorized before they fly to it and harvest it.

A mistaken flight to an alien colony could prove lethal—the guard bees there would make short work of any foreigners emitting the wrong signals.

Experienced foragers are more successful than recruits; in the course of their lives they have gathered not only honey but also information. This is possibly also the reason for the brain growth that has been ascertained. Experience stimulates stronger links between the nerve cells in certain areas of the brain that are particularly important for memory formation and thus increase the volume of the brain.

Menzel describes this process—the storing of information in the memory—in a characteristically scientific way as "alterations to the circuitry of the nerve cells." Yet in addition to all the definitions and dissections, the weights and measurements, and all the objective analyses, he adds, "You can make hypotheses, you can verify them or reject them but if the bee wasn't such a wonderful work of art, my urge to research wouldn't have got me too far. You have to love them both as individuals and as a social community."

Evolutionary biologists go into raptures when talking about bee family units. They regard a bee colony as an entire organism, like a highly structured body with many cells and organs. And the organs in this superorganism, to stick with our image, can change their functions. In the course of their lives, bees perform a variety of very different tasks.

A newly hatched bee, in the warm half of the year, begins its roughly five-week-long life as a cell cleaner; it cleans the brood cells that its younger sister-bees have just left. It only holds this job for one to two days. From the third until the twelfth day of its life it is a nurse bee, meaning that it feeds the larvae. For the first three days of life the larvae receive royal jelly, a juice produced in its salivary glands; then they receive normal rations of nectar and pollen. The queen is the only bee to be fed lifelong supplies of royal jelly— hence its name.

While the majority of bees hold a number of jobs during their lives, a few specialist courtiers remain to feed the queen. When the queen is fed, it secretes a substance that attracts these courtiers and ensures that they continue to do their jobs as royal providers. For the majority of bees, however, after a short time as cleaners and nurses there follows a period when they take on three special jobs one after the other.

First, they become specialists in processing the reserves of nectar; the foraging bees deliver it, the current crop of specialist bees chew it, and then fill and seal the cells. Simultaneously or subsequently, they qualify as honeycomb builders, also producing the building materials. The honeycomb is made from wax which they excrete from special glands on their abdomens. Wax is an incredible substance, consisting of around two hundred different compounds of saturated and unsaturated hydrocarbons, acids, and esters, to name but three components. Later, in their third development stage, they become guard bees at the entrance to the hive. Only those bees with the right family "smells" are granted entrance; every now and then, foreign drones that have strayed after a nuptial flight are also allowed in.

Only after twenty days—after they have hatched into bees, after they have been cleaner and nurse bees, and, according to the needs of the colony, after they have completed the three stages of training—do the honeybees fly off as foragers to deliver nectar, pollen, and water to the hive.

Water is transported to the hive in their honey stomachs, mainly to control the temperature in the hive, as it evaporates through a collective buzzing of wings protecting the sensitive brood from overheating. Within the hive a constant temperature of 35°C (95°F) has to be maintained, and only in the winter can the temperatures drop below 30°C (86°F). If heating is necessary, the bees decouple their wings so that their muscles run at full power without moving their wings.

In addition to the normal career sequence from nurse to load carrier, there is one more temporary, special development. In early fall, a certain generation of bees can considerably extend its normal five-week lifespan. These winter bees, as they are known, survive the winter by keeping the queen warm and feeding it, and can live in the hive for up to five months. Science is still seeking the mechanism that allows the organisms of bees to switch from a short lifespan to longevity at the end of the reproduction season. One hypothesis suggests that in fall the workers feed each other with royal jelly, the energizer that allows the queen to live so long. The theory is that as the production of eggs slows to a standstill with the onset of winter, royal jelly is no longer needed in the same quantities for feeding the queen and the larvae. This switch to longevity, albeit gradual, is still not completely understood.

The capabilities and endurance of bees is impressive, but possibly even more fascinating is the interplay. The question of what governs this is not only interesting to zoologists. For a number of years scientists have been focusing on social insects, and there have been increasing attempts at distilling the entirety of bee or ant colonies as conceptual models for human coexistence. "Global society can be seen as an autopoietic network of self-producing components, and therefore as a living system or superorganism." So said the Belgian star of theoretical cybernetics Francis Heylighen, who also coined the term "collective intelligence."[4] As with all biological processes, such superorganisms develop in the course of evolution and we are gradually beginning to under-stand more. Technologically, according to the principles of optimization, there have already been tangible results—the building of small robots, emulating the principles of swarms. Technicians have already succeeded in building small helicopters that can swarm like a flock of starlings without crashing into each other and artificial beetles that can overcome obstacles by linking together to form a bridge. And there are already drones that can fly in swarms, named after their archetypes in nature and made airworthy by funds from the US military budget.

Humans would not be humans if at this point they did not think ahead. Could we not make ourselves independent of bees? Is it unimaginable that pollination could be carried out by robot bees that can exchange information as the bees do? Opinions are divided as to whether this vision of the world, in which small, automated bees buzz around instead of the originals, is a scenario of hope or of horror.

CUSTOMIZED
BEES

EFORE WE DOMESTICATED bees a couple of thousand years ago, we encountered them in our role as hunter-gatherers. Although the point was not to hunt wild bees, gathering their honey was somehow linked to the dangers of hunting. Then we progressed to cultivating the land and rearing cattle, and also began to establish bee colonies near our dwellings. The oldest examples of beekeeping are probably from the Middle East; in Europe, beekeeping first really took off in the Middle Ages. It is difficult to say how and when the breeding of bees first had an impact on humans or even which bees were preferred. But these tended animals with a sting in their tails could be neither tethered nor confined when honey was harvested, so it is fairly safe to presume that beekeepers early on dreamed of how nice it would be if their honey producers could be a bit more cooperative.

A lot of things to do with beekeeping are unique—they cannot be transferred to other animals and cannot be compared to the breeding of other livestock. After all, a queen is not a huge mammal but a small insect, who in the course of its life has not half a dozen descendants (like a cow) or hundreds (like a sow) but millions of them. But still, a number of breeding principles apply, irrespective of the animal group.

In the professional rearing of bees, the term means more than just the breeding of offspring, it means controlled propagation in pursuit of quite specific breeding goals of genetic remodeling. Following the principles of genetic evaluation, the qualities that are nearest to our requirements should be targeted and reinforced by crossbreeding selected individuals and less favorable attributes should be suppressed accordingly.

The qualities that the beekeepers maintain or wish to improve can vary from country to country or from bee region to bee region. Three particular features, however, are consistent; honey yields, manageability, and health.

The honey yields are about the gathering capacities of a colony, about how much honey can be registered in one season—a busy bee is a good bee. Here, food consumption is significant: How much honey do the bees themselves consume or, as the case may be, how much supplementary foodstuff do they need? In this respect, the food source to brood graph presents an interesting picture. Ideally, a colony has its highest number and thus its strongest workforce when blossoms are peaking.

As far as manageability is concerned, bees get good marks for gentleness, consistency in the honeycomb, and a reluctance to swarm (the natural urge of bees to propagate by swarming should be minimal). They should react stoically and submissively to human encroachments and never defend aggressively. When a beekeeper accesses their living and working quarters they should remain peacefully on the honeycombs, on the frames where they have made their cells, and not leave them when they are removed from the hive or placed somewhere else for a short time. Beekeepers are also interested in how a colony handles propolis—the resin from buds that bees use as disinfectant in some areas of the hive and that is considered a nuisance as the frames become sticky and are difficult to move.

The third feature, the health of the bees, is mainly about resistance to diseases. Unfortunately, we know most about this factor when things are already bad and a disease or a parasite has already become established. This is becoming increasingly significant due to the current worldwide threat to honeybees. The resistance of a colony very much depends on its capacity for hygiene and, depending on the region, on the hardships of winter—how successful is the colony in overcoming up to six long, blossomless months of cold in the hive?

Evaluating the potential outcomes of breeding decisions is not a straightforward matter. The most important single factors are reckoned according to a point system, but there is no regional uniformity. Sound judgment and experience are called for in breeding evaluation as the factors cannot be precisely measured.

This is clearly exemplified by the assessment of hygiene. Swiss bee expert Ruedi Ritter explains:

> This impulse is important for the health of the colony. The colonies with the best marks are those that quickly clear the hive floor of debris and dead bees after the bad weather period. However, it is difficult to judge the cleaning instinct independently of the size of the colony because only a strong colony can warm the whole brood area during colder periods and in doing so keep the floor clean. Weaker colonies leave the dead bees and debris in the cooler, peripheral areas of the hive.[1]

In other words, as far as protection from germs is concerned, a numerically weak colony that cleans diligently is topped by a medium-strength colony of sloppy cleaners.

Uniform evaluation is also complicated by the fact that what is welcomed in some places is less appreciated elsewhere, if at all. So, rapid spring development—the ability of a colony to be ready to fly quickly and in significant numbers after the winter break—is most welcome at sites where early flowering plants need to be pollinated. This includes all rosaceous plants like apples, pears, cherries, and almonds. In areas where the food sources are in forests, it is better when a colony reaches full strength relatively late, in early summer.

They have to be later starters, because the aphids need to have developed enough for their secretions of honeydew to be harvested by the bees and processed into dark, strong honeydew honey.

For breeders who wish to optimize the qualities and talents of their animals, the question is: Which qualities are hereditary? That is, which ones can be influenced at all? Over the years, bee experts have developed a scale for "heritability," ranging from 0.0 (not hereditable) to 1.0 (entirely heritable). For example, a hereditary estimation could indicate a low value of 0.26 for honey yields, an average 0.41 for gentleness, and an impressive 0.91 for consistency in the honeycomb. But as the word "estimation" implies, although the values are given numbers they cannot be exactly defined mathematically.

Why the industriousness of *Apis mellifera*, of all things, cannot be genetically improved is, the experts say, to do with the fact that nature, over millions of years, has already carried out optimization through selection. Better foragers were always the enemies of the good foragers and they in turn were at some stage replaced by excellent ones. But as with top athletes, at a certain stage of perfection, a system, a creature, an organ can only be improved minimally because further progressions can only be made at the expense of losses elsewhere—which could cost nature dear.

It is easier to explain why the margins for gentleness are greater for beekeepers. Before we domesticated honeybees there was no selection pressure for placidity and humans could attempt to re-route priorities according to their needs. But as with all attempts at manipulation, as with all adjustments and readjustments, it is worth keeping an eye on the impact of mechanisms; a number of pieces of genetic information are often located at one control point. Dog breeders determined to breed Dalmatians with smaller black spots know now that this was only possible at the cost of impairing the dogs' hearing. It seems that the genetic factors that influence coat coloring are also associated with the development of hearing.

Breeders who select bees for winter hardiness, maybe to help them cope with northern winters, have to take into account a slight reduction in the colony strength. The advantage of a tolerance to the cold can only be achieved with this side effect. But because in colder climes spring arrives later, a well-populated colony would be burdened by a poor supply of blossoms anyway; for beekeepers in harsher climates, a moderate reduction in the size of a colony is not a damaging factor but really rather a positive one. Elsewhere there is a win-win situation for beekeepers. Tended colonies, in particular, inherit a low propensity to swarm, a factor highly rated by beekeepers.

How one attribute is encouraged while another is suppressed, how benefits can be developed without having to deal with significant drawbacks is the magic formula of beekeepers. There is both knowledge and a great amount of pseudo-knowledge. The beehive is always open to a world of esotericism and charlatans.

At the moment, great efforts are being geared toward *Varroa*-resistant or at least *Varroa*-tolerant honeybees. In densely populated regions of the world, purchasing the *Varroa*-hardened strains of Africanized bees—the "killer bees"—for breeding purposes is forbidden. They are easily provoked and aggressive, so the consequences would be unacceptable.

How one attribute is encouraged while another is suppressed, how benefits can be developed without having to deal with significant drawbacks is the magic formula of beekeepers.

What about other species of bees that over thousands of years have developed behavioral patterns that enable them to live with the mites? What about the Asian honeybees? It is well known that there is a species barrier between the Asian honeybee (*Apis cerana*) and the European honeybees (*Apis mellifera*) that cannot be crossed. Despite this, researchers are interested in learning what exactly it is that allows the *cerana* to co-exist with *Varroa*. It is striking, if not decisive, that the Asian worker bee broods develop in only eighteen days as opposed to the twenty-one days that the European ones need, or even twenty-four days until the actual hatching of the Asian drones. The *ceranas'* shorter development time makes it more difficult for the *Varroa* offspring, which thrive in the brood, to keep pace.

Can the European bees be induced to open the cell lids quicker? Not to date. As already stated, crossbreeding the species isn't an option. A genetic transfer of the *cerana* attributes would only be possible through genetic manipulation—if at all. Astonishingly, some committed environmentalists and beekeepers, who protest against agricultural genetic engineering, are willing to turn a blind eye to this modern magic if it could somehow help their bees.

Tentative steps that don't transgress the species' boundaries seem more promising. We know that bees that clean thoroughly make life difficult for the *Varroa* mites, and plans are already afoot to make very good cleaners from just good cleaners. There has been some success in observing cleaning bees with special infrared cameras and marking the particularly industrious ones. But how is it possible to pass on their traits when the workers in a colony are normally sterile females?

There is one exceptional circumstance in which the workers, to a limited extent, can reproduce, and that is when a colony loses the queen and the young brood, from which a new queen could develop, is not available. In this state of emergency a number of drones—laying worker bees—begin to lay eggs. As they are unfertilized eggs, only drones hatch from them and the colony will sooner or later perish, but their genes could still survive by being passed on to drones that leave the hive to mate.

This situation was artificially induced in the case of the best cleaner bees under infrared light. The drones laid eggs as laying worker bees. The hatched drones, genetically identical clones of workers, inherited the cleaning gene of their mothers, and the female offspring of the queen, fertilized with their sperm, really did prove to be better cleaners.

While the goals for the breeding of queens have changed and continue to change, the basics and techniques for doing so have remained fixed for a long time. Nature only allows a certain amount of scope for human intervention, and those wishing to understand how controlling intervening can be must know what the bee colony does when left alone.

In natural circumstances a colony will only rear a new queen when it needs to increase its population through splitting or if the queen is getting too old. It knows exactly when the time is ripe for this and carries out specific arrangements. Special brood cells, round and cup-shaped rather than hexagonal, are prepared for the young queens. The eggs that the queen lays there are the same as other fertilized eggs, which later become workers, and in the early days the larvae develop identically. Only the subsequent diet of royal jelly makes the difference, eventually leading to the hatching of fertile, female reproductive creatures.

But before this happens, at the earliest when the first new queen cell is sealed, the pre-empting swarm moves on with the old queen. Roughly a week later, further swarms, each with a number of competing future queens, also begin swarming. After the strongest new queen has asserted itself in a swarm and a new habitation has been found, it makes a nuptial flight and starts establishing a new colony.

In the rare event of "efficient" queen supersedure—that is, when the old queen gives in to the new queen without a fight—there is no swarming. The old and ailing queen remains in the hive and waits until the incoming queen mates. Once the young queen has begun to lay eggs, the deposed queen leaves the colony of its own accord to die. In conventional beekeeping, however, very few queens reach an age in which this form of colony-initiated regeneration can occur. Just as ways are found to stop the bees from swarming, it is a widely accepted beekeeping practice to kill the queen before its laying capacities begin to wane. Those who don't wish to let their bees decide when to propagate but still want to make two colonies from one, split them and artificially produce the loss of a queen in one half of the colony. Then the workers of this offshoot colony seize the initiative and produce a queen from larvae that would normally have been other workers. After hatching, the queen goes on a nuptial flight. Alternatively, the colony gets a foreign "ready-made" queen, previously mated, from a breeder.

In order to breed queens you need a cultivated colony with the best queen, fertilized with the sperm of the best drones, and one or more carer colonies without queens but with nurse bees.

When the first fertilized eggs containing the genetic information of a particular queen in the breeding colony become larvae, delicate manual intrusion is required: grafting. Equipped with delicate instruments, a clockmaker-sized

magnifying glass, and maybe a headlamp, the beekeeper removes very young female larvae from the worker cells and implants every single one separately in queen cup cells either hand-made from wax or bought versions made from polystyrene or other substances. Whether Swiss- or Chinese-style grafting tools are used, how much sugar syrup is subsequently dabbed onto the tiny larvae, and a multitude of other details are the stuff of lessons given to the novice by the bee master.

The pre-filled queen cups, usually a couple of dozen, are then hung between brood combs in a queenless nursing colony. In order for a nursing colony to be enticed to care for the cultivated larvae that have been introduced from outside of the colony, it has to be without a queen, and so the queen has to be removed. In this situation breeders can directly intervene. A colony in which the queen is missing does everything it can to end this threatening situation. When a queen has gone, the tone is audibly different, as if the bees realize that in a month the colony will perish if they don't produce a substitute quickly. By taking away the young brood you prevent the nurse bees from feeding their own larvae to fertility on royal jelly and instead have them accept the foreign larvae that has been imposed upon them. When they adopt newcomers with their artificial queen cups they construct the typical queen brood cells, which look like coarsely crocheted thimbles, around them and they are pampered to become future queens.

In addition to the genetic provenance of the implanted larvae, the key to the success or failure of the breeders' efforts to find the best queen lies in the strength and health of the nursing colony. The customer has the choice of having the queen impregnated by selected drones or introducing it to the colony as a virgin queen.

In the summer, everywhere that honeybees are kept, groups of drones await the arrival of a queen bee that will allow a few of them to mate. The drones can have different backgrounds, can be from separate colonies of various beekeepers, and can belong to diverse races. Those who opt for this form of mating leave to chance which of the neighborhood drones mate with the freshly hatched queen on its nuptial flight and consequently the genetic makeup of the offspring.

Breeders, and also ambitious beekeepers, select a different path at this stage and have their queens mated at remote locations, usually islands or secluded mountain regions where the breeders' interests in the mating partners of their queens are easier to control. Breeder associations and national agricultural organizations have set up "male hostels" with potential maters in such areas. These colonies have particularly high numbers of drones. These drones, according to the promises made to customers, have the desired good characteristics laid down in their genetic makeup. Beekeepers, sometimes from far away, arrive with their freshly hatched and still unfertilized queens, complete with a small entourage, and place their unmated queens in mini beehives in the vicinity of the "superdrones" and their supporting colonies. It is essential that the guests arrive without any drones and are only accompanied by worker bees. At these hostels it is almost impossible for unwanted competitors to interfere with the specially selected sperm donors. If all goes well, the beekeepers can return home with advantageously fertilized queens that have already begun laying eggs.

For those who want to take this a stage further, breeders can mail a queen that has already been fertilized by specially selected drones together with roughly ten nurse bees in attendance in a finger-sized, living animal transport cage. One end of the cage is sealed with a three-centimeter-thick (slightly thicker than an inch) plug of sugar mass; as soon as the new queen has been introduced to its new colony, the sugar plug is gnawed away from both sides. In the process the new queen and its attendants begin to take on the smell of the hive and what would otherwise have been a barrier to acceptance in a new colony then disappears.

As with horse and cattle breeding, there are top "studs" from which the owners promise miracles. Just as a horse-breeder might say, "My foal was sired by the wonder stallion Totilas," a *carnica* beekeeper might say, "My colony comes from a real Singer queen."

The Singers in the Ötschtal, Austria, are a true beekeeping dynasty. It seems somehow fitting that the family business is run today by two women, mother Liane and daughter Heidrun, as the queens from their breeding center are renowned. They have even registered Carnica-Singer, the race name together with their own surname, as a brand name. Under this name they offer a whole range of queens that can be ordered online and exported throughout the world. A pure-bred Carnica-Singer queen is priced at 60 euros (just over US$80), a previously mated high-quality Carnica-Singer breeding queen with provisions for the journey and attendants packed in special cases is on its way to you for 390 euros. A non-mated queen bee is to be had for just 16 euros and Singer worker bees sold as artificial swarms and *Varroa*-free bee material as a starter kit, for 38 euros per kilogram (almost US$24 per pound).

However, looking at the German honeybee scene as a whole, the majority of beekeepers continue to have their queens mated at their home locations. Comparatively few work with queens from professional pure-bred breeding centers, and although successful breeders are highly regarded, doubts are increasing about breeding approaches that narrow the gene pool, sometimes absurdly, even using outward appearances as a selection factor.

Ultimately, the breeders are working against the bees' own endeavors to avoid inbreeding. By nature, honeybees when mating try to affect a broad genetic mixture. Their drones fly quickly and far afield to spread their genetic makeup to other colonies. At a height of thirty meters (ninety feet) on every summer day between 1:30 PM and 4 PM, male bees from every possible direction meet at drone gathering areas to mate with queens from other colonies. They can then return with impunity to foreign colonies where they are fed and assimilated, something that the breeders would consider to be a disastrous scenario. People wanting to rear pure-bred queens would have to prevent these natural behavioral patterns.

Almost a hundred years ago, one breeder consciously and decisively pursued a different course, a course away from pure-breeding and toward a mixture of diverse hereditary factors. Karl Kehrle, born in 1898 in Mittelbiberach, Germany, was a pioneer, initially experiencing the loneliness of pioneering but eventually gaining widespread recognition.

He left his Upper Swabian home at a young age and while still in his teens joined the Benedictine monastery at Buckfast Abbey in Devon, England, where practical activities were encouraged on top of spiritual practices. The choice of activity was easy for Kehrle, as he had already fostered an interest in bees while in Germany. From 1919, under the name of Brother Adam, he took care of beekeeping. Southern England at that time had a major problem with serious bee plagues. The Isle of Wight disease, which we now know was a tracheal disease in which the mite *Acarapis woodi* penetrates the respiratory tracts, had wiped out almost all the country's colonies, including those of the monastery. The dark European honeybee *Apis mellifera mellifera* vanished from England.

Brother Adam was at a loss, but he did have a hunch about where to look for a remedy. A German "bee professor" named Ludwig Armbruster (1886–1973) had recently expressed well-founded, scientific doubts about the unconditional acceptance of pure-breeding—doubts that cost Armbruster his lectureship and university career and saw him denounced for "favoring foreign species" and "friendliness to Jews" when the Nazis took power. Armbruster steadfastly refused to distance himself from Jewish colleagues and their studies. His main denunciator, openly supportive of the Nazi regime, was rewarded with a step up the career ladder—an ascent that continued even after 1945, as he climbed toward the top of the German professional beekeeping hierarchy.

Armbruster used Mendel's laws of inheritance, which had been known for a while but were still not generally accepted, and applied them to bees in particular. His findings on the damaging effects of narrow pure-breeding and the importance of variety are, to a large extent, accepted today but are in no way followed in practice by all breeders.

Brother Adam, who all his life described himself as a student of Armbruster, set off to explore all the Mediterranean countries. *Auf der Suche nach den besten Bienenstämmen* (published in English as *In Search of the Best Strains of Bees*[2]) is the title of his scientific travelogue, which he dedicated to Armbruster. It became a classic in modern bee literature.

The Benedictine monk studied strains of bees in France, Switzerland, Austria, Italy, and Germany in 1950. In 1952 he turned his attentions to Algeria, Israel, and Jordan, and then to Cyprus, mainland Greece, and Crete; he also visited the Balkan countries and the Ligurian Alps, where he was particularly meticulous. In 1962 he was off again. This time his itinerary included Morocco, Turkey, Egypt, various Aegean Islands, and the Libyan Desert. Almost in passing he was able to prove that *sahariensis* was an independent type of bee, something that had been fiercely debated previously.

The practical result of his efforts—the gathering together of all the desirable bee traits—was a "new" bee which Kehrle named the Buckfast bee, in honor of his home monastery. He followed the crossbreeding methods of nature where two strains of bees impinge on each other geographically: the experimental mix. The Buckfast breeding practice, as it is still dynamically practiced, is based on the principle of continually crossbreeding with new strains. This process allows a pure-bred Buckfast line to be regularly replenished by the offspring of a parallel breeding line, with the parallel line consisting of a cross of Buckfast bees and various strains of European honey-bees. Almost more important than what Brother Adam did is what he didn't do. He wasn't interested in the physical features of the various races, like hairiness, coloring, or the vein structures of the wings. He was only interested in characteristics beyond appearance.

The Buckfast bees were no flash in the pan. In England, Kehrle's adopted country, and also in other corners of the world, the Buckfast bees proved to be diligent foragers with a fairly strong constitution. Walter Haefeker, who works with Buckfast bees, however, warns about the creation of the bee-loving Brother being the be-all and end-all: "These bees only prove their worth in the hands of beekeepers who provide enough food, be it from moving locations or feeding in good time after harvesting the honey. If this is neglected, Buckfast bee colonies, with their high fertility rates, are in danger."

Brother Adam's credo, proclaimed in countless lectures and publications, is as follows: Create as broad a genetic basis as possible and maintain it! Particularly in times of faltering colonies worldwide, this could be an important precept if it comes down to strengthening the immune system of the superorganism bee colony. As the virologist, cancer researcher, and bee authority Professor Eberhard Bengsch says, "Germs are nothing, the immune system is everything!"

Nowadays, since Kehrle's death in 1996, the success of the Buckfast bees is in the hands of the few breeders who use targeted mating to deliver time and again the special vitality required by migratory beekeepers. Despite their benefits—and the various local strains combine a number of economically viable traits—the Buckfast bees are not as extensively prevalent beyond the British Isles as the *carnica* in Central Europe. The Adam bees are held in disrepute by many breeders and beekeepers. In Austria, where they are not considered a strain but mongrels, they are even officially forbidden, one of the main charges against them being that an undesirably strong brooding impulse results from the mating of *carnica* queens with Buckfast drones. Those Buckfast beekeepers that remain between Lake Constance and Lake Neusiedl on the Austrian-Hungarian borders report of threatening anonymous letters, poisoned stocks, and burned-down hives.

Brother Adam's credo, proclaimed in countless lectures and publications, is as follows: Create as broad a genetic basis as possible and maintain it!

Whether *carnica, nigra,* or Buckfast bees, honeybees on the whole fend for themselves. They are livestock that don't have to be enclosed or herded, their hives take up little space, they don't leave dung heaps behind, and for six months of the year they even provide nutrition without having to be fed. Nevertheless, the workaday life of a beekeeper, even if part-time or as a hobby, does require commitment, consistency, and nowadays also a certain understanding of chemistry to cope responsibly with the evaporation of acids.

As with every agricultural activity, beekeeping is a year-long project with bouts of intense activity and more sedate phases, but never periods of total inactivity. What has to be done is decided by the bees, the season, and the weather. Depending on the weather or maybe dramatic events (for example, the outbreak of a disease), the beekeepers' annual program may be delayed or may intensify.

At the beginning of the calendar year all is quiet in the bee colony. In January, a gentle, uniform buzzing can be heard from the hive. In early February, but at most sites somewhat earlier, the queen starts to lay the first eggs. Depending on the state of the bee population, this month is still a quiet period for beekeepers. The bees clean their winter quarters and their bodies. At temperatures above 10°C (50°F) and in favorable weather conditions the cleansing flights begin and they defecate, something that they never do in the confines of the hive. All around the hives are flecks of honey-colored snow. The first pollens (hazelnut, alder, crocus, and snowdrops) and some nectar can be gathered. If the winter was unusually mild the beekeeper will begin the monitoring process: Is the colony faltering? Are there enough supplies available? How serious are the *Varroa* infestations?

March is brooding time in the colony. Considerable amounts of food are brought back to the hive to build up the colony. The growing colony needs new combs for both the nectar and the brood. Left to their own devices, bees would make their own combs, but beekeepers now supply either empty combs from the previous year, complete with the risk that they may be carrying germs, or they accelerate the building of new combs with comb foundation sheets embossed with cell forms.

In April, young bees, both workers and drones, are reared. With good weather conditions the early blossoms arrive (apples, pears, and cherries). It is still necessary to check whether the colony has enough materials for the building and filling of the combs.

From May, the delivery of pollen to the hive is in full swing. Within the colony the need to reproduce is awakened. The bees prepare queen cells for rearing new queens. The time is right for expanding the hives by placing new frames in the hive. Beekeepers who don't enjoy climbing trees to retrieve swarms would be well advised to run regular swarm checks and remove queen cells. Countermeasures like decreasing the colony by manual splitting can also reduce the risks of swarming. From the end of May until mid-June, depending on the region, the first harvesting of honey takes place.

In June, the development of the colony has reached its peak, possibly comprising 500,000-plus individuals. Work for the beekeeper in June is similar to the work in May. The spring honey can be extracted, and for those engaged in the pollination business, June is the prime time for migratory beekeeping.

After the swarming times are over in July, peace and quiet returns to the colony. The bees harvest early summer supplies (raspberries, blackberries, lime trees, and dog roses). And the beekeepers also harvest—the summer honey! At heights above one thousand meters (three thousand feet) and in areas with harsher climates there is, as a rule, only one honey harvest per year, in high summer or slightly later. Beekeepers wanting to produce honeydew honey relocate their colonies to nearby forests.

From August into September the colonies prepare for winter. The queen lays fewer eggs, and the production of offspring diminishes noticeably. The bees now hatching will be winter bees and will live not for five weeks but for six months. Drones are driven out of the hives. The only pollen entering the hive is from late bloomers; building activities cease. After the last honey harvest, beekeepers take on the struggle against *Varroa*, with ecological beekeepers using formic acid. As the nectar inputs decline in late summer the then unemployed foraging bees are used by some beekeepers to make artificial swarms to increase the bee populations. Particularly strong colonies are split, and the colony without a queen either gets a pre-purchased one or a home-bred one; if needs be, two weaker colonies can be merged into one. After a cool summer it is possible that even at the end of August, once there are no more blossoms, additional feeding may be necessary. October is, from a beekeeping perspective, a month of leisure. The queen has almost completely stopped laying eggs.

In November and December, the bees settle in for winter. The bees form a cluster in the hive, and from time to time the bees on the outside move toward the center to warm up. During the bees' active phase, the temperature is kept at a constant 35°C (95°F). In winter, the bees at the center of the cluster keep the temperature in the hive at 25°C–28°C (77°F–82°F) by shivering—they vibrate their flight muscles but keep their wings still, raising their body temperatures. The queen has priority, and as VIP of the bees, resides in the middle of the cluster. Beekeepers can make use of the dark months for further education and training. There is time to attend beekeeping lectures, for reading, and for improving equipment and accessories. At the Christmas markets, honey and other beekeeping products like wax candles can be sold. All that needs to be done with the hive is to keep a wary eye out for *Varroa*, using whatever treatment is considered necessary.

October is, from a beekeeping perspective, a month of leisure. The queen has almost completely stopped laying eggs.

If you ask beekeepers what the highlight of the year is, you won't get the same answer from everyone everywhere, but noticeably often the reply is that it is the moment after extraction when the kind and quality of the year's honey is revealed. Honey harvesting begins with an act of violence: Someone steals from the bees all that they have stockpiled for their feeding requirements, including their winter stores. But while in the olden days the harvesting of honey went hand in hand with the destruction of the colony, nowadays removal of the easy-to-handle frames holding the honeycomb has a comparatively minor impact. Before getting down to the sticky stuff, the beekeeper checks whether the honey is "ripe." This happens when the honeycombs are sealed or the honey in the non-sealed comb is so thick that even when the frame is banged against something solid nothing seeps out. Some people like to work scientifically and use a refractometer to measure the water content, which should not exceed 18 percent.

Bees remaining on the comb once the frames have been extracted can be dusted off with a bee brush. In modern beekeeping businesses, the brood combs are separate from the honeycombs so that the bees can be blasted from the honey supers (part of the commercial beehive) by compressed air—which is effective but can damage the wings—or by using repellants to drive the bees away from the honeycombs.

There are a variety of tools and techniques for uncapping, or removing the wax seal of the honeycomb. There is cold uncapping using a broad fork-like tool, and also warm uncapping with a spatula tool that has been warmed before use. Many beekeepers use a heat gun, which looks like a mini hairdryer. The uncapped honeycombs are then placed in an extractor, a centrifuge that slowly builds up speed and forces the honey out of the hexagonal wax cells. The honey accumulates at the bottom of the extractor and is drained into a honey tank where it remains sealed for a couple of days at room temperature. During this time, wax debris, which separated from the frames during the spinning process, floats to the surface where it can be easily skimmed off. Once the honey is free from particles it can be decanted into jars. In some honey regions, such as the northern Lüneburg Heide (often referred to in English as Lüneburg Heath) where there are still vast connected areas of heathland, portions of honeycomb are sold as a specialty product.

In addition to honey and wax, which bees secrete from special wax glands to build the honeycomb, bees produce other products. In specialized beekeeping businesses, the queen's diet of royal jelly is wrested from them and uses have been found for propolis, the substance that bees use as disinfectant. Propolis is a resinous mixture that bees collect from tree buds by taking the sticky cobwebby mass from the buds with their mandibles. Secretions that are released by the mandible glands in the process make the mass supple and the bees then stick it to pollen baskets on their back legs. Back at the hive, the gatherer remains almost motionless until its sister-bees have gnawed off the propolis-laden sacs.

Bees cover practically the whole honeycomb with a fine layer of propolis, and according to the classic bee text *Das Schweizerischen Bienenvater,* "They put propolis at the entrance like a doormat so that every arriving and departing bee comes into contact with it."[3]

At the time of the pharaohs, propolis served as an embalming agent. The ancient Egyptians possibly followed the lead of the bees. Small honey-robbing intruders that break into the hive and die there are coated in propolis, thus protecting the bees from the unwanted byproducts of decomposition that could contaminate their nest. Humans use the pleasant-smelling resin-like substance, which aids the healing of wounds, in ointments and creams.

China is the world's leading supplier of propolis, with annual yields of three hundred metric tons (just over 330 short tons)—an immense amount if you consider that a colony annually produces only 50–150 grams (two to five ounces). Very good foraging colonies like the Caucasian bees, however, can yield between 250 grams (8.8 ounces) and one kilogram (2.2 pounds). Depending on the region of origin, propolis can have a wide range of colors, from amber to anthracite. And somewhere hidden in this sticky substance—and yet to reveal itself—is the explanation for its marvelous sound qualities. Famous instrument makers such as Stradivari and Amati mixed propolis into their varnishes, but which kind, how exactly, and in what ratio to other substances remains their secret. The Singers, in Purgstall on the River Erlauf in Austria, not only breed queens but also sell other bee products—after all, the honey comes from an unspoiled nature reserve, free from pesticides and car fumes. The storage of the honey there well reflects the value of this precious natural product. In the cellar beneath the extraction room, the honey is stacked by the barrel, like liquid gold reserves in a safe.

5

HUMANS
AS BEES

WHILE IN THE USA the large-scale transportation of millions and millions of bees from one plantation to the next is impressive, there is another, very different method of pollination to marvel at in the People's Republic of China. In some parts of the country, two-legged pollinators with cotton swabs, brushes, homemade tools, and ancient medicine flasks filled with pollen abound. Something that we would consider almost unimaginable is very much a reality in parts of China: fruit trees that are pollinated by hand. Bees have not been made redundant because we have discovered better ways to do their job, but in some places, there simply are no bees to pollinate the apple trees. People wanting to harvest fruit in such places have to think of other ways to pollinate.

To see the human pollinators at work, Markus Imhoof and his film team traveled to Maoxian at apple-blossoming time. This was no easy venture, particularly for visitors from the West; the roads and bridges leading to the mountainous valley that winds its way to Tibet were ravaged during the devastating earthquake of 2008 and, to all intents and purposes, it is out of bounds to foreigners. The filming team of *More Than Honey* did, however, succeed in traveling to the high valley and were shown on-site how humans using their own materials copied a process that nature had perfectly equipped insects to perform.

The technique is as easy as it is painstaking. The pollen is collected by rubbing two blossoms against each other and catching the falling pollen in a newspaper; the pollen is then carefully dabbed onto each individual blossom. Improvised tools are used for this, mostly cigarette filters stuck onto a pencil or small bundles of chicken down attached to a stick. Every tree has to be visited and dabbed a number of times as the blossoms of one tree are never receptive all at the same time.

This lack of synchronicity is natural and serves a purpose. As a rule, the male and female organs of a blossom are close to each other, which could lead to self-pollination and limited genetic variability in future generations. In order to avoid the dangers of inbreeding depression, as this is called, and along with it a decline in resistance or fertility, pollen and the receptive stigmata develop at different times. So, when a bee brushes against ripe pollen in a blossom and then against the stigma of the same blossom, there are no consequences as the stigma is not yet receptive. However, the majority of stigmata in the neighborhood are receptive at this time.

The image of not worker bees but human workers going from blossom to blossom pollinating in the Sichuan province presents a horrifying glimpse of a future when bees may no longer be around. The painstaking work of the Chinese fruit farmers is the result of a situation that has nothing at all to do with bee mortality. The practice of using humans as pollinators has been widespread and well known for some time, as the anatomy of the small and most common honeybee in all of China, *Apis cerana*, does not suit all species of apples that have been cultivated in Maoxian since time immemorial.

This is why, even in the times of Mao Zedong, pollen was transferred by hand. The process was apparently documented in a contemporary movie sequence showing a young couple in the branches of adjacent fruit trees in full bloom, singing a love song while white blossoms swirl against the blue heavens. Since then, the ministry of agriculture has decreed that the Japanese Fuji apple trees should replace all the trees that require hand-pollination in apple plantations. The blossoms of the Fuji trees have shorter pistils and can easily be pollinated by the native Asian honeybees.

Markus Imhoof met a number of Chinese experts who had more confidence in hand-pollination than in insect-pollination. Are they right? As part of a state-aided project at the Agricultural University in Beijing, Professor Shi Wei researched and compared the pollination capacities of bees, bumblebees, and humans, something she had previously done in Kenya and Sweden. Humans landed far behind both types of bee in third place.

Bumblebees achieved second place most of the time but first place for some species. People interested in growing tomatoes on a large scale in greenhouses buy industrially produced bumblebee nests and hang them in the greenhouses. Companies offer nest boxes that can accommodate up to three colonies. Time-controlled lock-in systems using valves at the flight openings allow bees to enter but not to leave as there are some occasions, for instance when insecticides are being used, when the bees have to be kept safely in the hive. However, when compared with honeybees, the bumblebees do not fare as well in pollinating most blossoms. Strawberries, where each individual blossom has to be visited up to six times and with a specially choreographed circular landing, are very often deformed if pollinated by bumblebees; the same applies to apple and pear trees, which then tend to bear deformed fruits.

On the whole, honeybees retain first position in the pollination rankings, but whether there will be enough healthy colonies to maintain this ranking is uncertain, even in China. Scientists there who spoke to Markus Imhoof off-camera blamed the massive intrusions of pesticides for the dearth of bees in many of the fruit-farming areas of the huge country.

Hand-pollination has become so professional in some regions of China that the growing demand for pollen has led to the creation of a number of small businesses dealing in pollen. People like Zhang Zhao Su, from Wafangdian in northeastern China, make a living by harvesting, purchasing, transporting, and selling pollen. In her home in northern China there are traditionally large apple plantations but no longer bees, as most of them have fallen victim to pesticides or taken flight to escape them. Because of the harsh climate in the north and the very short blossoming time—only five days—fruit farmers rely on pollen from provinces two thousand kilometers (1,242 miles) to the south where the trees blossom six weeks earlier. This pollen, however, is not brought by bees but by Zhang Zhao Su, who travels with her small team to Taiyuan in Shanxi province in spring to buy apple blossom in bulk.

> It takes two days and two nights to drive there with two drivers taking it in turns to drive. As soon as we arrive we rent a large room where we can dry out the blossoms. When we have enough we start to process them. We cut off the pollen, the individual processing stages are, of course, a trade secret. Then we head back north. We have a fridge in the car which we plug into the mains when we stop for a rest, that way we are sure that the pollen stays fertile. When trees here in the north start blossoming we sell the pollen to fruit farmers.

It would be wasteful for the human propagators to use the concentrated pollen as bees do, so the pollen is cut with cornmeal, also offered for sale by the pollen dealer. Zhang Zhao Su packed her wares into five-gram (0.17 ounces) bags and sells them to retailers for CNY5 (roughly US$0.80). These are displayed right next to the agricultural toxins that are causing the bees all their problems.

Pesticides are a massive threat to bees in both the East and the West. The Asiatic *Apis cerana*, unlike the European *Apis mellifera*, copes very well with that great adversary of the honeybee, the dreaded *Varroa* mite. The smaller relatives of the European honeybees that have developed since the last Ice Age have a number of defense mechanisms against mite infestations at their disposal. Unlike the European strain, the bees from East Asia can smell when

their brood has been infested with mite eggs. They seal off the cells of infected offspring with such thick lids that the young bees cannot hatch and so die together with the *Varroa* mites. In the course of evolution they have devised a kind of euthanasia for the collective health benefits.

Varroa destructors have only appeared in Europe in recent decades. To find out how this mite was able to conquer almost the whole world we have to follow the tracks of the European honeybee, which spread around the world as a result of European emigration over several centuries.

At the end of the eighteenth century, a tsarist officer named Arshenevsky asked his sister in Kiev to organize the transport of twenty-four colonies of Ukrainian bees to his post, at that time in eastern Kazakhstan. They survived the four-month journey but succumbed to the Central Asian climate. The uniformed "bee fancier" then acquired colonies from the Urals. This relocation proved a success, and the bees, a strain of European honeybees, spread eastward in subsequent years via the Baikal region (1851), finally reaching the Pacific coast. The Trans-Siberian railway (1904) provided a link between East Asia and the western side of the huge Russian Empire and with it many Ukrainian emigrants—and their honeybees. The European strain, of course, came into contact with the native Asiatic honeybees that in past millennia had managed to learn to live with the tenacious parasitic mite. *Apis mellifera*, on the other hand, had no chance; they encountered *Varroa destructors* unprepared. The European honeybees simply hadn't had the time to develop strategies against what was for them an unknown parasite.

But there was also a second important distribution route, this one from the opposite direction. After European settlers had established *Apis mellifera* in America in the fifteenth century, the bees eventually reached Japan in 1876, where they met up with a subspecies of Asian honeybees that carried a special subspecies of *Varroa* mites. The newcomers naturally became infected. At the beginning of the 1970s, Japanese emigrants in turn brought their bees, now infected by *Varroa*, back to America. The infection spread from Paraguay to the whole South American continent and within a few years had reached the US west coast.

The mites arrived in Europe in the 1960s as European beekeepers imported Siberian bees whose honey production was said to be far superior to that of European bees. This soon proved to be a fallacy, but the flourishing business in Siberian bees was unstoppable. The imported bees, together with their parasites, had long found their way to European beekeepers.

The mite was first described in Bulgaria in 1967; four years later, *Varroa* appeared in Czechoslovakia, as it then was, in 1976. By 1977 it was in Germany; ten years later it had made its way back to the US by air, where a short time later it met up again with its infected relatives from Japan.[1]

Varroa is considered to be one of the main causes of the continuing mass mortality of honeybees in Europe and the USA, and, as the film team of *More Than Honey* discovered, the Chinese beekeepers who work with *Apis cerana*—bees that can coexist with *Varroa*—feel compelled to keep the mite away from their hives with the help of formic acid. Although *cerana* are able to fight the destructors on their own, Chinese breeders find that the hives are freed from parasites more quickly with the help of formic acid.

While researching the documentary in southern Sichuan, Markus Imhoof also met migratory beekeepers selling their honeydew honey on the side of a busy road. They openly admitted that they no longer place their hives in areas of intensive farming south of the mountains. The losses caused by the vast amounts of sprayed pesticides were simply too great. It is, so to speak, a form of industrial action by beekeeper and bee; the consequence is that in the meantime people have to pollinate the trees.

As Imhoof was about to buy some of the migratory beekeepers' wares, he was warned by his Chinese driver that he shouldn't be fooled by the idyllic site with the forest as a backdrop, as Chinese honey generally contains many harmful substances and, particularly in this region, is mixed with sugar syrup.

In the EU, the importing of Chinese honey was prohibited in 2002. The concentrations of chloramphenicol were too high, which suggests that high dosages of antibiotics were being used to counter American foulbrood. The ban was lifted in 2004 after the Chinese prohibited producers from using antibiotics, but the bad image of honey from East Asia remained and was reflected in market misgivings.

There were and are ways for the Chinese honey exporters' products to reach the supermarket shelves. The standard method is blending, meaning that contaminated honey is mixed with other kinds of honey. The act of blending is legal and even appears on the labels: "A blend of honeys from EU and non-EU countries."

In order to sneak honey of dubious quality or of allegedly suspicious origin through the EU security barriers, the suppliers have to eliminate traces of pollen or elements of pollen, which usually remain in the end product. The specific composition of these elements provides information about whether a honey (or its diverse blended compounds) originates from Argentina, South Africa, the Lüneburg Heide, or even China. For imported honey, there are no binding lower limits to the amounts of pollen that a tested honey batch has to contain. When the inspectors detect noticeably low levels of pollen, they may suspect that someone is deliberately trying to hide the country of origin—erasing the "fingerprint" by filtering out the pollen—but nothing more. They make their judgment, suspicions notwithstanding, only on the measurable pollens. If, for example, one-third of a batch of honey consists of high-grade Hungarian acacia honey with a natural proportion of pollen, the other two-thirds being global market honey with barely detectable amounts of pollen, the test report will then identify it as honey from Hungary. The unusually low pollen content of the honey blend clearly points to adulteration but the findings of the pollen analysis stand. And they indicate that, in this case, the honey is from an unproblematic part of the world.

For critics and experts in this area, it is very difficult to imagine that such a loophole in the testing procedure is not the result of specific lobbying. Walter Haefeker commented that:

> There is no way that I want to say that imported honey is under a blanket of suspicion, after all 80 percent of the honey that we eat in Germany comes from other countries, but I have always been a bit concerned about what the lab specialists from the testing institutes tell me about being unable to find the countries of origin after analyzing pollen from large amounts of imported honey. And honey producers from India and China proudly declare that they produce "honey" that has absolutely no connection to a single bee but is made from various types of inverted sugar. And bear in mind that they were talking about honey that they had successfully guided past the EU inspections.

Not only does China produce the bulk of honey consumed globally, but it is also the leading worldwide producer of royal jelly. Markus Imhoof, in the run-up to filming, visited a factory near Hangzhou that produces 30 percent of Chinese-produced honey and 70 percent of its royal jelly. The modern building in which the royal jelly produced drop by drop by nurse bees is processed looks like a high-security wing. This is hardly surprising as it harbors a substance that is both precious and correlatively expensive, and that has to be shielded from any form of contamination with none of it wasted. If we consider the laborious production process, it is all the more astonishing to see the size and number of stainless steel drums into which the substance is poured and stirred by workers who are dressed and masked like surgeons in an operating room.

In order to maximize yields of royal jelly, the manufacturers have found ways of making the bees produce far more of the precious substance than they would usually do in the hive. If one colony had several queens laying eggs instead of a single queen as found in nature, the output could be increased many times over. Normally a queen doesn't tolerate any rivals in its colony—it kills them. In the Chinese royal jelly factory they cut off a mandible, as the queen can only use its lethal sting after having firmly gripped its rival with its mandibles. After such treatment, seventy queens can lay about two thousand eggs apiece in one hive, giving a daily total of almost 10 million eggs.

In the first three days after hatching, the larvae are transferred to artificial queen cells and subsequently placed in a nursing colony from which the queen has been removed. In a bee colony, the absence of a queen constitutes an acute emergency—without the egg-laying queen a colony would die within a few weeks. The nursing colony now feeds not only one queen larva but also all the artificially implanted queen larvae. A robust nursing colony can be expected to support fifty feeding areas. Once the maximum royal jelly production has been reached in the hive after a few days, the young queen larvae are removed and the valuable larval food taken from the cell. A strong nursing colony can produce half a kilogram (around a pound) of royal jelly per season.

Normally a queen doesn't tolerate any rivals in its colony— it kills them.

Royal jelly is particularly valued in traditional Chinese medicine for its supposed life-prolonging properties. Its reputation can probably be traced to the observation that queens, which eat nothing but royal jelly, can live to the age of eight while the other bees die after five weeks. Whether you can prolong your own life by consuming it is questionable, but at prices of up to 130 euros per kilogram (almost US$82 per pound) you are likely to die impoverished.

The scope of its possible applications sounds impressive. Royal jelly preparations apparently increase mental and physical capacities, and followers of alternative medicine in particular swear by it. Even the enterprising cosmetic industry has discovered the substance.

The royal jelly is a profitable sideline, although it's almost impossible to confirm the figures. China, which apparently makes 90 percent of the royal jelly available on the market, produces between 2,500 and 3,000 metric tons (2,755–3,306 short tons) annually. With world market prices between 100 and 130 euros per kilogram (around US$63 to US$82 per pound) this means a volume of trade of up to 390 million euros (US$540 million) per year for these global market leaders.

But it is not only royal jelly that has fascinating properties. Apitherapy, which involves using bee products for medicinal purposes, has had some reported successes. Manuka honey from the blossoms of the manuka myrtle, *Leptospermum scoparium*, originating from New Zealand, is enjoying increasing popularity. The Maoris have long known of the disinfecting properties of this honey and long used it externally for wound healing. In 2011, manuka honey, under the brand name of Medihoney, was approved as a medical product. Even though most bacteria cannot grow in honey, this honey is additionally sterilized by gamma irradiation. Detailed studies about the effectiveness of manuka honey have yet to be completed, but it is already certain that this honey has far higher levels of methylglyoxal than other available honeys. Methylglyoxal is formed in the honeycomb as a sugar degradation product and has been shown to have antibacterial effects. This bacteria-killing potential, which is also present in food-grade honey, allows the bees to store the viscous mass in the hive in the first place.

Whether for its healing effects or its tastiness, honey remains a very special sweetener—long after the discovery of how sugar can be manufactured from sugar cane, or more recently from sugar beet. You could even claim that it is a substance with cult potential.

While globally bees are becoming less common in the regions shaped by agriculture, beekeepers are experiencing a renaissance in cities of all places.

For a number of years, metropolitan areas like London, New York, and Tokyo—famous for skyscrapers, smog, and noise but not at first glance the ideal refuge for bees and other insects—have been finding increasing numbers of supporters of a phenomenon called urban beekeeping.

It all started in Paris. In the mid-1980s, a retired opera house prop man named Jean Paucton began placing beehives on the roof of the Palais Garnier opera house. Initially it was just a small, private indulgence that brought in a good additional income. Paucton sold small jars of his Opera Honey for 4 euros (around US$5.50) to the gift shop in the foyer of the opera house where they sold for 14.50 euros. The bees found nectar in the abundance of blossoms in the city parks, from window boxes on balconies, and from linden trees and acacias, yielding some 450 kilograms (992 pounds) a year. His experiment received a lot of attention after the publication of photographs by *Earth from Above* photographer Yann Arthus-Bertrand in the popular magazine *Paris Match*. Since then, Paucton has been inundated with requests from photographers and camera teams, journalists, and groups of visitors from around the globe.

Today he uses this interest to campaign not only for his particular form of urban beekeeping but also for the general good of bees and a nontoxic environment. He claims that his losses within the colonies amount to roughly 5 percent, whereas traditional French beekeepers in the countryside have to bear losses of almost 50 percent. He places the blame firmly on conventional agriculture: "There just aren't any proper farmers anymore, there are only agricultural companies and they all use pesticides."[2] Paucton sees his success, as measured in honey yields and the health of his bees, as indirect proof that agricultural toxins are a main cause of the worldwide death of bees because, at least in this respect, cities are as good as free from toxins. Walter Haefeker thoroughly agrees with him: "In the meantime, bees in the cities are much better off than those in the country."

One of the reasons for this is the biodiversity that has now become a feature of urban areas. There are apparently more species of plants thriving in Berlin than in all the agricultural areas of Germany. "The city has become a

While globally bees
are becoming less common
in the regions shaped by
agriculture, beekeepers are
experiencing a renaissance
in cities of all places.

mixed forest," says the evolutionary biologist Josef Reichholf. "Even alluvial forests with the richest biodiversity can't even begin to compete with the diversity of trees in large cities."[3] This might be explained by the fact that alluvial forests are characterized by tree species that can adapt to varying water levels, whereas trees in urban environments have no need to adapt to dramatic variations in water levels. And gardens—at least for the enlightened modern gardener—are no longer products of uptight adherence to planning, with austere right angles and featureless lawns, but instead offer more diversity and are planned more around the needs of fauna.

In the cities, the annual average temperatures are higher by some 2°C (an increase of 3.6°F), which also helps the bees. The warmer climate means that the foraging season is somewhat longer; in addition, the winters are milder. Concerns that the honey from urban areas could be contaminated by exhaust fumes and fine particle pollution have proven unfounded. Bees always collect fresh nectar that has not had time to come into contact with air pollutants and absorb them. Should any of the mostly fat-soluble pollutants somehow reach the hive, they bind to the wax in the honeycomb and so have no effect on the honey.

The continuing trend of beekeeping in the cities has attracted such an increase in interest that beekeeping initiatives like Berlin summt! (which translates literally as Berlin buzzes) are at the limits of their training capacities—beekeeping has to be learned. Just like the Parisian experiment, the initiative has been behind colonies being placed on prestigious Berlin buildings, such as the Abgeordnetenhaus (the state parliament building), the Haus der Kulturen der Welt (house of the cultures of the world), Berliner Dom (Berlin cathedral), and many other buildings to publicize its cause. And in Vienna, Austria, bees launch off the roofs of the Burgtheater, the Vienna State Opera, and since 2012, the gilded dome of the Vienna Secession.

There is, however, a development in the other direction that at first glance seems to be a paradox. While the number of beekeepers has begun to rise in recent years, there are ever fewer colonies. In Germany, according to the German beekeeping association, the number has dropped from around

850,000 colonies at the beginning of the twenty-first century to 700,000 today, a decrease of almost 20 percent in just over a decade.

Fewer colonies despite more beekeepers? This is less of a contradiction than it seems, as urban beekeepers seldom keep more than three colonies. Interest in beekeeping mostly translates as completing one's own natural garden, experiencing the fascinating world of bees, and attempting to redress the deficit of bees in the meadows in and around cities. Modern beekeeping seems to be more of an expression of ecological awareness and an urban yearning for nature than a lucrative sideline or a gentle pursuit for the retired, which was often the case in the nineteenth and twentieth centuries.

The relocation to metropolitan areas has natural consequences for the pollination situation in the countryside. Multitudes of small colonies belonging to the new hobby beekeepers live and forage behind houses, and in gardens, parks, and cemeteries.

In built-up areas the bees are not lacking anything, but the open countryside and farmlands, where ambitious recreational beekeepers placed their hives in the good old days, have gradually offered fewer species of plants and provisions have become scarcer. Both toxins and blossom uniformity affect bees and other insects not only in the long-term but also, in some cases, immediately. So-called harvest shock can destroy whole colonies within a short period of time. When the remaining, still fertile meadows are mown at a single stroke—providers of agricultural services have to make their expensive equipment economically viable—the supply situation inside the hive becomes precarious within hours. Sometimes an entire rural district is mown in one day. The bees react promptly and drastically; developing broods are torn from their cells so that the population of the hive reflects the new and reduced food supply situation. If, however, a hive is numerically reduced in this way, the intruders, the *Varroa* mites, quickly gain the upper hand and the decimated colony can no longer withstand them. In addition to this, thousands of insects are direct victims of the high-performance machines that squeeze liquids from freshly mown grasses to shorten drying times and thus contribute to the disappearance of bees. According to a study from the Schweiz. Zentrum für Bienenforschung (the Swiss bee research center), between 35 percent and 53 percent of the bees active in fields during the harvest die as a result of these farming methods.

Furthermore, the increasing demands for biofuels and electricity from biogas plants require ever larger areas for cultivation; the meadows and fallow lands on which bees and other insects rely as food sources are being swallowed up by the industrialization of agriculture. This caused the German professional beekeeping association to launch a publicity campaign in the summer of 2012. Operating under the name Flower Power (they opted to use English for their name), and using sponsorship, bonus systems, and targeted subsidies, the beekeeping association hoped to persuade biogas plant owners to grow bee-friendly flowering plants and to convert these plants into energy instead of growing corn, which is causing increasingly more damage to the soil and the countryside than can be compensated for by biopower, for this purpose. The biogas industry has, to say the least, an image problem because of the fiercely deplored, featureless tracts of land taken up by corn cultivation nationwide and so maybe will be open to suggestions. Blossoming landscapes *and* power from the fields? Now that's a sweet idea.

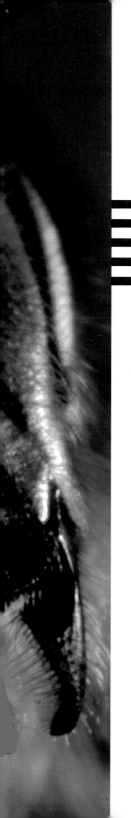

BEES: AN
UNTAMED FORCE

N 1956, THE zoologist and bee researcher Warwick Estevam Kerr imported 120 *Apis mellifera scutellata* queen bees from South Africa to Brazil, where the predominant strain until then had been the Italian honeybee, *Apis mellifera ligustica*. In a laboratory 130 kilometers (almost eighty-one miles) west of Rio de Janeiro, he and his team began experimenting with crossbreeding the two strains. They could not have foreseen that the results of their endeavors would one day gain infamy throughout the world as "killer bees."

The background to Kerr's experiments was a contract from the Brazilian government to look for a strain of bees that could better cope with the local climate conditions than the imported species of European honeybees.

Until the fifteenth century, *Apis mellifera ligustica* was unknown in both North and South America. The first immigrants and settlers brought the bees with them by ship to the New World. The western honeybee strains, used to mild continental or Mediterranean conditions, "worked" in the New World, but in tropical and subtropical zones were much less productive than in their native habitats.

Five hundred years later, Warwick Kerr searched for solutions to resolve these issues by crossbreeding African bees, which are less sensitive to heat, with European ones. He hoped to combine the good characteristics—the gentleness of the European bee and the heat tolerance of the African bee—in one bee. The objective was a productive bee that could cope with the climate of South America.

Kerr was well aware that his bees were not to escape to the wild, but even test colonies cannot be kept in complete isolation. Bees are dependent on their workers' gathering nutrition in the form of nectar and pollen, and so to prevent the spread of new hybrid bees while allowing this natural requirement to gather, Kerr devised a special grid for the entrance holes that was just wide enough to allow the workers but not the slightly larger drones and queens through. This appeared to be safe enough as only these latter bees could breed. The grid would normally have prevented the test bees from gaining freedom had not one day a particularly animal-loving member of Kerr's staff removed the grids. He felt sorry for the foragers that, on returning to the hive, had to shed their pollen baskets in front of the small entrance holes.

Thirty-six queens swarmed out and formed their own colonies. Shortly afterward, their drones mated with the local *ligustica* and the uncontrolled crossbreeding of both races began. The offspring coped splendidly with the tropical climate, but to general dismay, the following generations failed to display the relative gentleness of the African *scutellata* and absolutely none of the marked gentleness of the European *ligustica* drones from whose genetic makeup they had been developed. Instead, the paradoxical result was mild + mild = wild.

The Africanized honeybees (AHB), as the new entity was soon named, defended their hives with a vigor that surprised everyone. And terrorized! With European bees, only the few guard bees attack potential enemies—whether they are humans or bears zeroing in on the bees' life insurance of honey supplies—but the Africanized bees follow raiders for a kilometer or more (more than half a mile) en masse. Thousands of years of domestication seemed to have been blown away—it was comparable to a phlegmatic European lap cat being crossed with a mild Egyptian one, with the result being a wild cat that not only arches its back and hisses but also instantly attacks with extended claws.

Kerr and his colleagues tried to recapture the bees and to poison them, but the speed at which their breed spread gave them no chance. At a rate of three hundred to five hundred kilometers (186–310 miles) per year, the Africanized bees conquered the continent, entered Central America, and from the 1980s onward began occupying the southern states of the USA. By 2005 the "killer bees" had claimed a seven-hundred-kilometer (435-mile) strip from southern California almost to the Mississippi, and have since reached Florida.

The Africanized honeybees (AHB), as the new entity was soon named, defended their hives with a vigor that surprised everyone. And terrorized!

The flying capability of the Africanized drones is a significant factor in their high rate of expansion. They start earlier in the afternoon and remain airborne for several hours longer than their European counterparts. On top of this, they fly faster, so they are more likely to be where they need to be when a sexually mature queen appears.

Kerr was a 30-year-old scientist when he accidentally became the father of the "bee disaster." He spent the majority of his time into old age explaining the background to the episode and trying to placate beekeepers, scientists, and the general public.

But he continued, even in Brazil, to try to introduce gentleness into the breed. In 2005, in an interview on the fiftieth anniversary of the "killer bee" Big

Bang, he explained why his efforts ended in failure: Brazilian beekeepers were simply not interested. They quickly became used to the fact that the Africanized bees produced sixty to eighty kilograms (132–176 pounds) of honey, much more than their European counterparts for whom fifty kilograms (110 pounds) is considered a very good yield.[1] Compared to this, the additional costs for better protective clothing and, if need be, more effective smoking equipment to fend off attacking bees were of little consequence.

Within a short period of time, Brazil was catapulted from nowhere to being the sixth-largest honey exporter after China, the USA, Argentina, Mexico, and Canada (2012). In the northeast of this huge country in particular, beekeeping became a significant economic factor in a region where previously the European bees had only produced modest returns. And so it seemed that the costs to society as a whole in the largest country in South America had suddenly been put into perspective. "Previously," said Kerr, "125 people a year died as a result of insects in Brazil, twenty-five from bees. From our bee breeding program the number increased to 195 bee deaths. [...] Alone in São Paolo on one particular road we have five times as many traffic deaths annually."[2]

Kerr remembers vividly the reversal of his image from Frankenbee breeder to savior: "At conventions up until the beginning of the 1970s, it was common for wives of beekeepers to point at me and tell their children: 'That's the evil man who made the killer bees!' At around 1974/75 this all changed suddenly and the women were saying: 'That's the man who rescued farming, thanks to him Dad can buy a new tractor, now go and say thank-you!'"[3]

In the southern US states that were "colonized" in the 1980s via Mexico, the "killer bees" rapidly gained the status of a hostile army. With or without a political subtext, fear, hysteria, and horror always promise good business and a new "killer bee" subgenre appeared in the horror movie genre. The Japanese movie *Genocide* (1968) predated the "bee invasion," as did *Terror Out of the Sky* (1978) and *The Swarm* (1978), with its impressive star-studded cast in which not only the bees but also Michael Caine and Henry Fonda were let loose on the public. The movie poster shows an oversized swarm of bees the shape and size of a gigantic tornado that threatens to overwhelm a typical American skyline. It was the time of the Cold War, and no cliché was too well worn to prepare Americans for attacks on the good guys (USA) by the realms of evil ("killer bees" or communists). Even years later this motif was still good, with such titles as *Deadly Invasion: The Killer Bee Nightmare* (1995), *Killer Bees!* (2002), and the German TV movie *Die Bienen—Tödliche Bedrohun* (released in English as *Killer Swarm*) (2008).

In news programs, "killer bee" attacks flickered on to the screen, with close-ups for millions of viewers at prime time penetrating the public perception more than forest fires or falling coconuts, both of which cause more fatalities in the USA than bee attacks.

Nevertheless, this did nothing to slow the demand for the high-performance honey producers. "Every year," says Kerr "beekeepers come from the US [to Brazil] wanting to buy our bees. We don't sell, and anyway importing is illegal." Markus Imhoof also experienced this contradiction—on the one hand, panic-stricken fear of "killer bees," and on the other, an open admiration of their abilities. For his bee documentary, he decided to focus his camera on the Africanized bees, asking himself if there wasn't maybe a ray of hope hidden in what was generally perceived as a threat.

If a biopic were made of Fred Terry, the beekeeper from the southern states whom the director encountered, then the "killer bees" would buzz the theme tune. Terry comes from southern Arizona and looks like Burt Lancaster's twin brother. He is charismatic, a good raconteur who hones his skills during frequent lectures and even more so as a country singer—a stroke of luck for a filmmaker looking for protagonists who can effortlessly and precisely get to the point.

Terry's history with "killer bees" had many phases. It began in the early 1990s as he helplessly watched his "good," domesticated European bees in his hometown of Oracle, Arizona, take a serious battering from bee diseases, and in particular, attacks from *Varroa* mites. At the same time he heard reports of "bad" bees way down south in Brazil, bees that were wild but astonishingly resistant to mites and other bee tormentors. He became interested. How bad could the bees be when they were good against all kinds of epidemics?

Terry traveled to Brazil, and to his astonishment he met beekeepers who successfully worked with Africanized bees in a relaxed way and without risks, and who thought that the fact that their bees were called "killer bees" in the USA was a bad joke. While he was in Mexico, he met one of these beekeepers in a taco bar. "While we were eating I was hinting at my sympathies for Africanized bees. When we were finished and had left the bar my colleague yelled at me: 'Man! Don't talk about Africanized bees when others are listening. I've got some and I don't want my neighbors to know! I don't want any problems!'" Obviously the conflict between humans and bees there still had to be resolved.

Almost a year after this encounter he read in a trade journal that the Mexican beekeeping convention had officially decided to view working with Africanized honey bees as standard practice. Terry made a decision: "I will stop fighting them. I will join them!"

For Terry, what finally tipped the scales to switching from European honeybees to Professor Kerr's accidental new breed was confirmation of a rumor that the number one scourge of bees, the *Varroa* mite, was unable to harm Africanized bees. This *Varroa*-tolerance was until then only known in the small Asiatic *cerana* bees, which somewhere in East Asia, in the process of evolution, had developed resistance while passing on the plague to the defenseless European honeybees.

Terry quickly became an active champion of what was generally being presented as a threat. The pro arguments, however, fell on deaf ears during the heated discussions on the "attacking death squads" in the USA in the 1980s and 1990s. The debate revolved solely around whether and how the "killer bees" could be stopped from crossing the border into the USA on a grand scale. There were discussions about, for instance, fine mesh nets and corridors saturated with pesticides to keep the area free of bees. The idea of making the Central American isthmus a no-go area for bees even made it to talk shows and reader forums.

Suggestions were made, all of them bordering on insanity, but the line dividing reason and madness became blurred in the face of floods of reports

The idea of making the Central American isthmus a no-go area for bees even made it to talk shows and reader forums.

about people being attacked and cattle stung to death. Running parallel to the reports on catastrophes, magazines and newspapers featured articles with tips on how to prevent the stings from taking your own lifeblood. There were even serious discussions on a nationwide inoculation campaign using an antiserum from the blood of beekeepers who are exposed to frequent stings.

Terry is still surprised to this day, thirty years after the first outbreak of "killer bee" hysteria: "As far as I know, in the last twenty-five years there have been eighteen fatalities from bee attacks and most of them elderly. Every year in the USA there are thirty thousand deaths from firearms, forty thousand die on the roads." But according to Terry there is little tolerance for such comparisons: "Among other things, it's because we Americans are terrified of being invaded. It's only happened once since the foundation of the United States, on March 9, 1916, as the Mexican revolutionary Pancho Villa with some hundred soldiers attacked Columbus, New Mexico, burned down the town and made off with military equipment. The invasion only lasted one day, but after all, Pancho Villa came from Mexico just like the killer bees."

Killer bee phobia even brought prosperity to a number of people. Terry says that some of the KKBees, as he refers to the killers of "killer bees," amassed millions of dollars by working as pest controllers who specialized in destroying swarms that had landed near human habitation or even in the rafters of houses. The recapturing of swarms, which beekeepers have dealt with for no pay for centuries, suddenly cost up to US$1,000 per assignment.

Prices collapsed as word spread at the end of the 1990s that the incarnation of evil wasn't really as evil as the bee busters in Terry's neighborhood claimed in their US$2,000–US$3,000 advertisements. Nevertheless, this new version of homeland security remains in business to this day and is still being touted online. One such provider uses a kind of Batman mask for his assignments and markets his honey as Killer Bee Honey.

Fred Terry does just the opposite. His honey jars are labeled "Desert Honey from the Singing Beekeeper." He doesn't kill the killers when he recaptures swarms, and he doesn't charge anything for gathering them from roofs, or garden sheds, or wherever else they might have settled. His payment is the bees themselves; he takes them home and works with them. He just doesn't broadcast it.

As to why the hysteria spread like a wildfire, Terry suggests a classic fear—the wild horde. "A swarm is always a good bogeyman." How real and immediate the danger actually is plays a minor role. Danger always contains a large portion of subjectivity. Terry loves relating the amusement that the beekeepers in Brazil and other South American countries get when told that their bees are called "killer bees" in the USA.

> Let's be honest, isn't it strange that we accept that there are many gunshot victims but nobody speaks of "killer guns." We accept that there are even more victims of car accidents, but nobody speaks of "killer cars." We don't call other insects like spiders or scorpion "killers." In the USA there is a wasp called yellow jacket that is credited with far more "murders" than the "killer bee." Nobody has suggested calling this wasp the "killer jacket."

Even Terry, who has only a few hives behind his house, the rest being exiled to the desert a safe distance away just for safety's sake, didn't tell anyone in the early years that he had been keeping "killer bees" since 1991: "We US citizens make up just 4.5 percent of the world population but 75 percent of all lawyers, most of them busy as bees and buzzing around chasing indemnity cases."

Before Terry, who is by no means a gambler, became an African bee guy, he gathered information from Brazil, a number of Central American countries, and Mexico and discovered that almost every accusation leveled against the "killer bees" was false. He never tires of emphasizing this.

It was said that not only did they not make good honey, they made too little of it. The exact opposite is true. It was said that their pollinating capacities left much to be desired. Wrong: They are in no way inferior to the European bees. And finally, it was said that you couldn't handle them safely. Almost wrong! It would be better to say that Africanized bees cannot be placed near human habitations or busy routes. And even with protective clothing and smoke, you still have to be more careful with them than with tended bees. Beekeeping in swimming trunks—a popular image of the European domestication ideal—wouldn't be a good idea, and even Terry looks more like a knight in armor when interacting with his bees. He advocates adapting to the bees and not forcing them into some pattern just for the sake of making things easier for humans. He also points out that once it is dark there are no risks involved in approaching the hives: "Honey-stealing bears have known this for ages. They come during the night when no bees fly, not even in emergencies."

And so Terry, when he is not busy with beekeeping or singing, has become a kind of duty counsel for the defense of the Africans, as he calls them. He finds the formal term, Africanized bees, too colonial.

After he has listed and disproved the standard arguments yet again, it mostly boils down to one thing that is difficult to explain because it's rooted somewhere deep in the collective subconscious: The bad guys are threatening to drive out the good guys. Or, bad beats good. To some extent, this is just another example of un-American activities.

But eventually cracks began to appear in the resistance of the general public, and Terry upped the ante: "Look at cattle: They can survive in dry and dusty Arizona. They are wild and healthy and not fancy, they can even eat cactus. Just imagine trying to get along here with big, fat Holsteins. No chance! Zero! The Africans are not pure bred, not domesticated and tamed; they've remained primal, wild, and resistant."

With the increasing danger of epidemics and the worsening supply of blossoms, the wild European honeybee swarms have virtually no chances of survival. That was once very different. In the sixteenth and seventeenth centuries, European settlers brought strains of *Apis mellifera ligustica, carnica, mellifera,* and *lamarckii* honeybees to North America. Swarms that escaped from beekeepers resettled very successfully in the various climate zones, even in semideserts, prairies, or the harsh mountain climates of the Rockies.

Nowadays, there are hardly any wild swarms of honeybees with a European background. Their status lies somewhere between "highly endangered" and "extinct." "The European bees are the losers and the Africanized the winners. If I was a gambler I would bet on their survival. The Africans will still be here when we no longer exist," Terry tells me with a smile, radiating calm and serenity almost as if the peaceful day were approaching when the planet Earth would be rid of *Homo sapiens.*

But before that happens, *Homo sapiens* (particularly in the form of *Homo scientificus*) will try everything possible to deliver the bees from their enemies the *Varroa destructors,* which continue to pose the biggest threat. At the moment, the parasitic mites are on the minds of a couple of thousand scientists worldwide who are working on containing them.

Bees affected by *Varroa* do not die from loss of blood to the blood-sucking mites, which go particularly for larvae but also attack adult bees, but from the secondary effects. Pathogens enter the hemolymph via the bite wounds and then the bloodstream. A typical and striking result of such infection is wing deformity. Workers bitten during the larva stage hatch with stunted wings, making them unfit to fly and condemning them to a short life.

Shortly before the sealing of the brood cells, fertilized female *Varroa* mites invade the cells and give birth to their offspring. These develop in the cells and tap into the hemolymphs of the larvae so that when the bee hatches, a whole mite family joins the colony. In this way, the *Varroa* population in a hive can double in just under one month.

And the *Varroa* of today can breed an extra generation of tormentors—possibly an effect of climate change, because the summers increasingly linger deep into fall. "One reproductive cycle more is really a critical factor!" says *Varroa* expert Dr. Eva Rademacher from Freie Universität, Berlin. Just how diseased a colony is can be measured by the number of *Varroa* mites living in a colony at a particular point of time in the bee year. Eva Rademacher's rule of thumb estimation—useful as long as it isn't used as the only indicator—involves the sticky board on the floor of the hive; the fallen bodies of dead mites remain stuck there and are relatively easy to count.

The degree and extent of damage to a colony is significantly influenced by the length of development of the bee brood; the longer an individual bee remains in the sealed brood cell the better it is for the parasitic lodger, although there are differences between male and female bees. Drones need

twenty-four days from egg to hatching, the female workers only twenty-one; likewise, drone larvae need 6.5 to 6.9 days for metamorphosis in the sealed cells and future workers only 5.5 to 6.2 days.

The extra day means that the parasites in the drone brood cells can develop better and in greater numbers. The mites seem to know about the small difference; lab tests at the end of the 1980s and early 1990s established that *Varroa destructors*, if left to their own devices, choose to infest drone broods. Scientists assume that fertile mites carried (involuntarily) by nurse bees receive information from somewhere about whether future drones or future female bees are inside the unsealed brood cells. This information could be a chemical signal, or perhaps the mites are able to recognize the slightly different construction of the cells—the surface of the drone cells is 1.7 times larger than that of the worker cells.

One tested and toxin-free defense strategy of beekeepers is drone cutting, which makes use of the behavioral patterns of the mites by destroying as many of the drone brood as possible before hatching. By doing this, future infestations can be contained. The missing males are not relevant for the survival of the colony.

Special trapping combs that use a similar principle have also proved effective. When there are no broods, for instance, when a newly established colony has yet to produce offspring, the mites are lured to a drone brood comb. The mites pounce on the only available resource and in doing so no longer threaten valuable parts of the bee population. The beekeeper then just has to choose the right time to dispose of the infected and now sealed trapping combs. "The process is highly effective. On average 87.7% of the mites were trapped in the 73 units that were treated," wrote the scientists Berg, Schmidt-Bailey, and Fuchs.[4]

As is so often the case in scientific and medical research into illnesses, the key question is: Why are some individuals or groups of individuals either not affected or able to live with an illness? How is it that the Africanized honeybees are able to cope so well with the pandemic?

A Brazilian team of researchers studied the higher *Varroa*-tolerance of the Africanized bees and were able to confirm in 2011 what had been obvious in

Latin America for years: "In tropical regions of Brazil, where honeybees are Africanized, the mite effect on the colonies appears to be reduced to the point where no control measures are necessary and no colony losses because of this mite seem to occur." And another remarkable finding:

> Artificial infestation of bee colonies with adult *Varroa* females in São Paolo, Brazil, has showed that Africanized bee workers were almost eight fold more efficient in getting rid of the mites on their bodies compared to pure Italian bee workers. Artificially infested Africanized bees reacted to the presence of *Varroa* from the very beginning of infestation. Strong body movements involving the abdomen, legs and mandibles were performed by infested workers. The movements executed by infested workers permitted nearby workers to identify the *Varroa* on their body. When fellow workers identified the presence of the mite on the body of another worker, they used their tongues and mandibles to attack it.[5]

Proficiency in recognition and removal of mites are not the only advantages. Africanized bees were also able to prove in a scientifically controlled study carried out in Mexico in 2002 that they displayed considerably better grooming behavior than European bees when removing infected broods from the hive.[6] The "killer bees" were considerably better plague preventers; the Italian bees only cleared away 8 percent of their infected broods while the Africanized bees managed to clear away 32.5 percent.[7]

These comparisons in themselves could provide part of the explanation as to how the descendants of Warwick E. Kerr's crossbreeding experiment were better suited to fighting the destructors than the previously dominant European honeybees. But there were still other indications. Various groups of scientists had noticed that in colonies with high resistance there were strikingly large numbers of infertile mites. In 1999, an astonishingly high proportion—43 percent—of infertile female mites was registered in "killer bee" hives; in European beehives, the proportion was only 19 percent.[8] In other words, in European beehives, more than double the number of mites were busy producing parasitic offspring. Unfortunately, more recent studies—in 2003—found that the mite infertility rates between Africanized and European bees were approximately matched.

Even if the Africanized bees do have numerous advantages, though, nobody would want to give them freedom of movement throughout Europe. "It's not only about beekeepers working without danger, it's also about the safety of joggers peacefully jogging past a nearby hive," says Walter Haefeker, appealing to all those in densely populated central Europe who are rather thoughtlessly recommending switching to Africanized bees because they are the best collectors. Hope in the fight against the mites rests at the moment with those offering and promising anti-*Varroa* weapons and strategies.

The battle against *Varroa* is increasingly becoming a matter of selection for beekeeping. It is hardly possible for hobby beekeepers to deal with mites unless they are completely indifferent to yields and losses. So, just as yields from monocultures of industrialized farming can no longer be obtained without insecticides, fungicides, and pesticides, beekeeping seems to have become impossible without some kind of permanent defense against *Varroa* and other diseases. At least, that is the opinion of the majority. The danger is that the beekeepers who don't engage in a systematic fight against the *Varroa* endanger other colonies in their neighborhood. Collapsing colonies from uncontrolled *Varroa* infestations infect healthy ones within a large radius.

Bee activity, measured in hatching offspring per working day, and *Varroa* infestation form two very different activity curves. In May/June, life in the hive reaches its zenith with up to fifty thousand bees in the hive. In contrast, the peak of mite density happens in October, when bee life is just ticking over and preparing for the winter rest phase. This is a critical moment, and the mites hit the colony most severely when it is beginning to slow down for winter.

Should the winter generation—the longer-living bees that overwinter with the queen—be severely weakened by mites in fall, the flame of life may flicker or even be snuffed out during the cold winter months. It is almost as if Amundsen were to set off in the extreme cold with a sick and undernourished team.

But a faltering colony is not only a danger to itself. Bees are no angels, and given the chance, they will plunder the food supplies of other, weakened hives. Along with the stolen honey, they bring the parasites of the dying colony back to their own hives. Beekeepers' responsibility to act against bee diseases goes beyond a responsibility to their own hives; they also bear some responsibility for the well-being of other colonies in the vicinity. Beekeepers making a stand against *Varroa* are encouraged by associations and research institutes to use combined defense systems and not to put all their trust in one single substance to provide the solution, even if its properties are extolled by the large chemical multinationals. Many mites have already developed resistance to substances contained in Apistan, Bayvarol, and Klatan, and critics of chemical treatments are worried that the multinationals have unintentionally but effectively created supermites through aggressive marketing and use of their products. However, nobody dares say this for fear of lawsuits.

Bees are no angels, and given the chance, they will plunder the food supplies of other, weakened hives.

Even as far as conventional beekeeping is concerned, before and during the foraging season for nectar and pollen, treatment with chemical substances should not be an option, proclaims the Arbeitsgemeinschaft der Institute für Bienenforschung e.V. (the Association of German Bee Research Institutes). It emphasizes this point with a bold exclamation mark.[9] Honey cannot afford to lose its reputation as a pure, natural product.

When the final honey harvest has been delivered, the workers continue to raise the brood. At the same time, the offspring are nurturing the mites. Methods of fighting *Varroa* are put into action in order to get at the source of evil after the honey harvest from high summer until October, if possible killing them in the cells or just after hatching. Formic acid is the substance of choice, but its application requires a high degree of skill, experience, and expertise. In winter, when there is nothing to gather and the queen has suspended the laying of eggs, beekeepers are advised to use lactic acid and oxalic acid, finely sprayed and following the recommended dosage. Too much weakens the bees, too little does no harm to the mites.

But on top of "too much" and "too little," is not a third "something else" possible? As Fred Terry fills his jars with light golden, runny semidesert honey from his "killer bees," he occasionally shares his thoughts, which are maybe too simple and too true to immediately make sense. "Why don't we just listen to the bees? Swarming bees fly away from an infested hive. They fly away from the mites and start all over again. Professional beekeepers prevent swarming and in doing so are leaving their colonies to the parasites." Terry guides a "killer bee" to a small opening in the window in his filling room and watches it fly off before he continues. "We've bred out all aggression from bees and maybe also weakened their resistance to all sorts of diseases. Now, suddenly there are bees that are resistant, but we are walking away from them instead of teaming up with them."

BEES OF
THE FUTURE

ORIGINALLY, AS IN America, there were no honeybees in Australia. European settlers brought their beehives to Australia with them by ship in 1832. Thanks to isolation and strict quarantine regulations, the descendants of these bees were still free from *Varroa* in August 2012. Despite this, beekeepers, governmental authorities, and bee experts are preparing for the time when this may no longer be the case. The country is in a state of siege and reports that individual swarms of wild Asian or European bees, potentially afflicted by *Varroa,* could reach the continent by ship or air are a source of concern throughout the continent.

In 2011 alone, twenty-five such colonies of bees were discovered by the Australian quarantine inspectors and fortunately intercepted in time. Even if just one colony infested by *Varroa* finds its way into the country, it would be enough to establish the destructive mite in Australia. The general fear is that Melbourne is the most likely gateway for *Varroa*, as its large harbor and airport and its proximity to already afflicted New Zealand make it especially vulnerable.

In 2007, we saw how quickly an uninvited guest could enter Australasia. At that time, a swarm of Asian bees, the main host of *Varroa*, managed to swarm on land, without being noticed, from the mast of a yacht. Fortunately, these invaders were not only later established to be free from *Varroa*, but were also discovered early and destroyed. Nevertheless, the public authorities had to declare that the Asian bees had become established in Australia at the end of 2011. The expectation is that the mites will arrive and will spread *Varroa* throughout the continent.

Professor Boris Baer, director of the Centre for Integrative Bee Research (CIBER) at the University of Western Australia in Perth, sees the delay in the mite reaching Australia as a unique chance. "We are working on reacting calmly and having a running start when the time comes. If all goes well, we will have alternatives to the usual barrage of chemicals," he said. Defense through chemical substances is only a temporary solution; parasites are adaptable and develop resistance to pesticides and antibiotics so that newer substances have to be continuously tried and tested before being applied. Additionally, as the parasites cannot be treated with these substances without affecting the host, conventional chemical treatment also carries with it the danger that the bees will be weakened and the destructors strengthened.

Even if just one colony infested by *Varroa* finds its way into the country, it would be enough to establish the destructive mite in Australia.

This means that alternatives have to be found. At CIBER, three interest groups—beekeepers/breeders, government authorities, and scientists, groups that normally keep well clear of each other—have joined forces. This group has been working together since 2008 and is financed by the Australian government, Western Australian bee breeders, and international research grants, and is all about using synergy "so that the right questions reach the right ears and good answers don't get lost somewhere amongst real or digital files," said Baer. He cites as an example a positive lesson from the recent past. Western Australian beekeepers and breeders have already withstood one plague, American foulbrood, which ravaged the country twenty years ago and which was said to be impossible to overcome without the use of chemicals. This experience should encourage them to approach the *Varroa* problem without relying on chemicals.

Once an Australian state was infected with American foulbrood, its borders with other states were closed to bees and bee products. Despite addressing the problem without using aggressive antibiotics, there were no catastrophic losses. "Local bee breeders are still able to cope reasonably well with foulbrood," says Baer. "Bees are bred, by and large, to manage the disease by themselves." Researchers at CIBER are counting on similar results when *Varroa* or other diseases eventually reach the continent. They have confidence in the powers of resistance of nature and are trying to strengthen them.

CIBER has not only brought together beekeepers, government officials, and scientists, its own research team consists of experts from various disciplines, ranging from biochemistry to classical biology and from molecular biology to physics. Normally, scientists dealing with "how" questions (How can we explain metabolic processes?) are not interested in "why" questions (Why do metabolic processes happen in such and such a way?). But here they approach the answers to a key question together. Professor Baer explains: "What is the molecular fingerprint of a bee with strong resistance? What is it that makes it different to a vulnerable bee? And even better, how can we share this with the honeybees that we have bred? Unique opportunities are emerging to fully understand biological processes."

One PhD student at CIBER discovered that the ejaculate of drones protects against certain infections as it contains special molecules that kill the unicellular fungus *Nosema apis*. Interestingly, the bees themselves have developed this form of resistance, a natural defense, as it were. The next step will be to identify the active agents that the bees have developed against these intestinal parasites and maybe use them as medication. Additionally, it would be interesting to test the ejaculate of drones from different colonies for *Nosema* resistance and later to breed them with bee strains that are particularly resistant.

The research team also suspects that the sperm from drones from hives that are *Varroa*-tolerant could be used to establish a strain that, in time, would also be *Varroa*-tolerant. In this respect they are also considering carrying out an experiment that could be controlled because it would be conducted on an isolated island off the Western Australian coast which at the moment is reserved for military forces. Under continuous scientific observation, native queens would be artificially inseminated by imported sperm from drones from *Varroa*-tolerant colonies and would raise new colonies there.

Who knows? Maybe the "new colonies" from this experiment could be mixed with *Varroa*-infected European honeybees in, say, Papua New Guinea in controlled field trials. If a strain of the island test bees developed into a *Varroa*-resistant strain and were then repatriated to Perth, they could function as a genetically programmed local weapon in the battle against the *Varroa destructor*.

Results of preliminary trials from 2011, in which a hundred Australian queens were exported to the USA to raise new colonies in *Varroa*-infected areas, have not discouraged the CIBER team. The US beekeepers, for the first time in their careers, discovered more mites than bees in their hives. The experiment failed, or to quote the wry comment of Professor Baer: "There's room for improvement."

For a number of years it has been recognized that parasites can be transmitted by bee sperm. If they were introduced to a new colony by artificial insemination, however, nothing would be gained from increased mite resistance; it would be replacing one evil with another.

Professor Baer sees another starting point in Australia's feral bees. Feral bees shouldn't be confused with the diverse types of bee that are generally termed "wild bees." Ferals are wild honeybee colonies that trace their ancestry to the bees that the European settlers brought to Australia in 1832. They escaped from human custody and survived in the wild where they developed a particularly good immune system unaided by outside help.

The widespread feral bees in Australia managed to regain an astonishing amount of what animals in the wild have to know. Wild bees have to find ways to tackle local diseases. As they are not kept in boxes or given extra rations in winter, only those colonies that pass on their genes despite infestations, winter temperatures, and drought are strong enough to produce young queens and drones.

The CIBER team is planning to bring back some of this capacity to resistance to the cultured Western Australian bees. The scientific breeding objectives are clearly defined by Baer:

> We have to go back, and in this case it is not a step back. We have continued to distance our honeybees from what a bee is and should be. Bees are no longer able to cope with the challenges of parasites, pesticides, and climatic changes as their immune systems have been ruined through breeding where the only interest is in X amount of kilos of honey per colony. And if *Varroa* doesn't get them then the next plague or the one after that will. We have to go back to go forward.

Evolution is the teacher, the pacemaker, and the signpost. Life remains on the planet after 4 billion years mainly because it has reinvented itself countless times. The question of how this was possible keeps both practical and theoretical evolutionary biologists busy.

Darwin created the foundations of the theory of evolution with his book *On the Origin of Species*. His theoretical concept stated that animals that are better adapted to their environment than their competitors pass on their genes because they reproduce more successfully and produce better adapted offspring. Under given conditions, species that are less well adapted are less capable of passing on their genes. They either become extinct or find a new niche better suited to their special abilities. So, for instance, plants that are crowded out by faster-growing species adapt to colder climates where their competitors are not able to compete. Natural selection, survival of the fittest, adaptability—today Darwin's keywords have become common knowledge. Thanks to numerous research projects, the empirical proof of their accuracy has been supported, even if sometimes they are quoted out of context and cited in terms that are not politically correct. An abridged understanding of evolution and development of the species has prevailed for a long time, at least in the popular imagination. A living creature continues to develop to secure its own survival, and thus the survival of its species. Cheetahs improve their sprinting speeds to be able to hunt down the particularly fast antelopes. Antelopes have to become faster to escape from the cheetahs. And all of this because in life there is an integrated master plan that demands that you have to improve to survive.

Nature demands from its creatures that their genetic information, their blueprint, be passed on to the next generation. Nature is pretty flexible about how exactly this should take place. The fine-tuning of certain individual qualities is not the only development principle. Individuals can also have their genetic information passed on instead of passing it on themselves. Although such behavior is widespread in the animal world, even Darwin recognized that the behavior of individuals that seemingly altruistically help others to reproduce was difficult to explain with his theory of evolution. With the bees, the workers do not, as a rule, lay the eggs, but they do help the queen to raise the offspring.

How can such behavior arise when genes are egoistic units that are only supposed to reproduce themselves? It took until 1962 before an evolutionary biologist named Bill Hamilton came up with the answer. He developed a mathematical model, the concept of inclusive fitness. According to Hamilton, an individual profits from the services of the helper if the individual that receives the services is related and thus shares the genes of the helper. In other words, an individual also propagates genes through relatives such as nephews or nieces. This concept explains altruistic behavior and negates the conflict with Darwin's theory.

The scientific community finally understood Hamilton's brilliant premise some twelve years after its publication, and further developed his ideas to form a branch of science that is known today as sociobiology. The term "kin

selection" from this discipline states that individuals from a certain species abandon their own reproduction and help related individuals to pass on shared genes to the next generation.

The principle is most pronounced in insect societies. Worker bees perform their grueling workload in the service of the community. Perhaps a slightly less exhausting schedule would enable them to live for a little longer than the usual five weeks that lie between the birth and death of a simple honeybee. But only egocentrics pose such questions. An individual bee has achieved its purpose when it has ensured the passing on of its genes as effectively as possible. The best way of achieving this in a colony is by supporting the queen, which shares at least 50 percent of the same genes and which can pass on the genetic building blocks not only one time but millions of times.

To imagine a colony of bees in its entirety as an organism is not so easy, but fundamentally a bee community acts as a single living entity. It consists of an asexual body (a multitude of workers performing various body functions), an organ responsible for conception and birth (the queen), and another organ that takes on fertilization (the drones). Biologists term such a combination of teamwork a "superorganism."

Superorganisms like ant, termite, or honeybee colonies act differently from individuals; their future depends on more than the ability for survival of individuals. For social insects, benefits are intended for the greater good. The strength of the species lies in the community. Had evolution, hundreds of millions of years ago (insect communities have existed for at least this long), only catered for the advantages of a single primeval insect, maybe even an ancestor of *Apis mellifera*, there wouldn't have been social communities.

There are examples of kinship among mammals as well as the socially organized insects. Young, pre-adult vixens assist their mothers in rearing younger offspring and so support the continuity of 50 percent of their genes—that is, the proportion that they share with the cubs. This is a good decision as long as the genes cannot be passed on directly via the birth of their own offspring. When a marmot is on lookout duties for predators it risks being caught by an eagle. Its altruism, however, protects its own genes, which are also found in the closely related clan in the vicinity. Greatly simplified, the kinship principle is *family first*. If you cannot pass on your genes effectively, support those that can.

In addition to the honeybees that live in colonies, the ones we are familiar with, there are other "homeless" bees. We currently know of twenty thousand types of bees divided into nine families, and there are probably more. They are spread throughout the world wherever there is blossom, everywhere except in Antarctica. In Germany, the term "wild bees" has become accepted, and at the moment around six hundred different types are just about surviving—"just" being the operative word that we have become used to when talking about diversity, and it characterizes the state of affairs of wild bees. According to the estimates of Bund für Umwelt und Naturschutz (BUND; Friends of the Earth, Germany), 80 percent of wild bees in Bavaria are endangered or even threatened with extinction.

Although there are no specific studies, we can assume that wild bees, just like tended bees, die from the effects of sprays used in agriculture. The main cause of their deaths, however, has to be the loss of habitat. Wild bee biotopes are disappearing; structurally rich forest margins, hedgerows, extensively farmed pasturelands—lightly fertilized or not at all—remote meadow orchards, small sandy areas and clay ridges and their associated plant life, field margins, and waterside areas have all become increasingly rare.

As a rule, wild bees depend on the diversity inherent to a natural landscape, and modern farming with its standardization and depletion of flora is killing them. The most threatened of the endangered species are bees that are dependent on a single kind of blossom. Or looking at it a from different angle, a small number of flowers are dependent on being fertilized by certain bee or insect species. The experts speak of this mutual dependence as a close symbiotic relationship. For example, a species of mining bees, *Andrena hattorfiana*, can only survive if the right mixture of *Knautia* (such as *Veronica*), *Scabiosa* (such as *Centaurea*), and the thistle-like *Dipsacus* are in the vicinity, and these are groups of flowers that have become scarce in our highly cultivated landscapes.

The fact that we seldom recognize or even notice wild bees as bees isn't only because they have become scarce and because their appearance, to the untrained eye, is similar to that of our honeybee. Most of them don't swarm and their combs are not found in hollow tree trunks—for the layman, two classic signs of bees. The majority of wild bees are solitary, loner bees.

The life cycle of these solitary bees is very different to that of the super-organism honeybees. While honeybees are specialists, the solitary wild bees could be labeled as generalists. Every fertilized female takes care of building the nest and raises its offspring alone. There is no parallel division of labor as with the honeybees. As a loner, a wild bee builds the brood cells, lays eggs, and forages for pollen and nectar, which is fed to the larvae as a more or less runny mixture. Finally, the solitary bee mother seals the brood cells in which the larvae go through their metamorphosis. Solitary bees can rear two generations in a year, and in especially warm years with an Indian summer, sometimes three. Much of what the honeybees have to do to organize their social life is not applicable to the wild bees. They don't have to stockpile and there is no need for the waggle dance to pass on information. Solitary wild bees don't have to take up a series of jobs in their lifetimes, which at times honeybees have to do exclusively. Everything that has to be done has to be done one thing at a time by the mother bee. Drones, the male offspring of wild bees, are, as with honeybees, little more than flying sexual organs; they leave the nest well fed, mate with female bees of their species, and then die.

Most wild bees have very short stings that cannot penetrate human skin. They have no need for the defense capacities that honeybees have as there is not much for wild bees to defend; as far as propagation goes, it is better to fly away and rear offspring elsewhere when attacked than to risk life. Solitary bees can also be high-performance pollinators: *Anthophora* can visit up to 8,800 blossoms a day, more than double the capacity of a single honeybee.

Entomologists have discovered that, in the realm of wild bees, there are degrees of solitariness and sociability, with some bees being more social than others. There are quasi-social bee types—the mothers performing their maternal duties next to each other, nursery to nursery as it were—and semi-social bee types—mothers that share responsibility for the brood, to some degree at least. Biologists recognize stages of evolution in the development of solitary bees—from primitive eusocials (for instance, colony-building bumblebees) to the eusocial honeybees.

The bumblebee occupies one such intermediate stage on the way to being a superorganism. Their colonies produce reproductive animals in fall. The drones and the non-mated queens from various colonies fly off and mate. The drones die in the process and the young, now fertilized, queens gorge on pollen and nectar and overwinter alone, often in deserted mouse nests from the end of August until mid-April. In spring they crawl out of their winter quarters and search first for early blossoms and then for a nest site where they begin to build the first wax cells, adding a bit of pollen and nectar and then laying their eggs. The queen rears the first workers until they are able to fly and take care of their mother, which can then concentrate on laying more eggs.

People use bumblebees commercially in greenhouses, very often for the pollination of tomatoes. These, like the blossoms of all the nightshade family, rely on buzz pollination. The insects vibrate their flight muscles rapidly to shake loose the pollen from the anthers. As species of the nightshade family do not produce nectar, the pollen is the only incentive for bumblebees or bees to visit the blossom, and it serves as nutrition for larvae because it contains particularly high proportions of protein and nitrogen. Honeybees do not get to the pollen by buzzing but by undoing the anthers to access the pollen.

Cuckoo bumblebees are an interesting subspecies that specialize in making others rear their brood. Just like a cuckoo profiting from the absence of a reed warbler or a wagtail to deposit its eggs in a host nest, cuckoo bumblebees lay their eggs in host brood cells, ingeniously choosing cells that have already been stocked for future larvae. The same law applies for these bumblebees as for all freeloaders: If they are too successful, they endanger the continued existence of their host. Their survival as a species is only secure if they make use of their form of outsourced parental care in moderation. Their relationship has become well balanced over an evolutionary period of time. The same applied originally to the relationship between the parasitic *Varroa* mite and the host bee. It became unbalanced only when another species of bee came into play so that not only the Asiatic *cerana* bees were afflicted but also the European honeybees. As we know, the encounter between *Varroa destructor* and *Apis meliferra* can be traced back to human interference. It is quite possible that at some stage they would have encountered each other anyway, but at a less dramatic rate.

The same law applies for these bumblebees as for all freeloaders: If they are too successful, they endanger the continued existence of their host.

The same goes for other factors that are making life difficult for honeybees and wild bees. Even creatures with short gaps between generations like insects struggle to adapt to the speed with which landscapes are being transformed by humans. And unlike birds and mammals, they have few food and brood reserves. At best they benefit from the protection of the more valuable habitats of species such as birds, amphibians, reptiles, and mammals that have captured the imagination of the public. Vertebrates are simply better suited as mascots of nature conservation.

But there are increasing signs in the media that conservationists also want to provide protective measures for six-legged creatures. The Internet is awash with numerous tips on which plants and nesting aids can be used to attract wild bees to parks and gardens. The tips, instructions, and offers range from complete bee hotels (ready-made in DIY stores) to simple pieces of wood with variously sized holes drilled in them and everything in between—nesting tiles, nesting boards, clay plates with or without perforations, reed boxes, tree disks, bundles of thin bamboo canes. And every now and then suggestions are made that require little effort, but might be about doing something that some gardeners don't find too easy: Gardens shouldn't be too organized and it helps to leave a few seminatural corners for guests like birds, hedgehogs (in Europe) or porcupines (in North America), butterflies, beetles, and wild bees to live, feed, and make their nests in.

In such places you can occasionally witness something special, such as the daily toil of the leafcutter bee described in detail by Karl von Frisch, the pioneer of bee research:

> The leafcutter bee makes a hole in rotten wood, then flies off to a green leaf of a rose or elderberry or a raspberry bush or something similar and then cuts with her sharp mandibles a slightly elongated circular piece out of the leaf, rolls it together and flies back to her burrowed tunnel, then she makes a thimble-like form out of the little piece of the leaf for a cradle. In the leafy thimble she places some food and then lays her egg on top, then she seals the opening with another circular piece of leaf.[1]

We don't know how many generations of insects it took for a simple field, forest, and meadow bee to become such an accomplished builder, but we do know that it could not have happened suddenly; it must have taken an unimaginably high number of intermediate stages. And each one of these stages must have had a purpose, otherwise evolution would not have allowed them to be genetically stable. We don't know the exact route to perfection; we only know that nature can allow a great deal of time. Since, however, humans have transformed the environmental conditions quicker than creatures can adapt, the time for development and adaptation seems to be running dramatically short. This is particularly true of creatures with long gaps between the generations, large animals like tigers, rhinos, gorillas, and birds. And maybe honeybees, too.

Pessimists could deduce from all this that our own future is not looking so good either, but we can view it a little less fatalistically. It doesn't seem improbable that the global honeybee population can survive the invasion of the *Varroa* if we support them. Hope doesn't lie in further chemical miracle cures but rather in approaches like the CIBER group's approach of isolating, producing, and applying measures against *Varroa* by using the natural immune response of bees.

As far as the sinister CCD is concerned, the lack of precise understanding makes it difficult to develop comparable concrete strategies. How are you supposed to combat an invisible enemy? Experts believe that this phenomenon is the result of a combination of different factors, one of which is collateral damage from chemical pesticides that produce a side effect of overriding the bees' orientation systems.

But whether the bees are being attacked by insecticides or mites, the bee colonies are more severely affected, as their immune response has already been weakened by unbalanced breeding efforts and by industrial approaches to beekeeping. We want as much honey and as many almonds, strawberries, etc., as possible as soon as possible—all for as little money as possible and with as few bee stings as possible. We have adapted our favored bees in the course of their domestication to these requirements, maybe a little too diligently. Maybe the example of the honeybee is once more showing us that the Earth cannot be subdued with impunity, particularly if this power is being exercised without limits and without respite.

Maybe it isn't too late to give some thoughts to how we could again adapt. One thing remains true: Nature can survive without humans but humans cannot survive without nature.

The Origins
of the Documentary

MORE THAN
HONEY

By Marcus Imhoof

BEES HAVE NOURISHED my family for more than a hundred years. They were part of my grand father's cannery business, Imhoof & Casserini, in Zofingen, a village in northern Switzerland. The cherries, apricots, raspberries, and gherkins would not have flourished in his vast fruit and berry garden without bees. "One-third of everything that we as humans eat wouldn't exist without them," my grand-father used to say.

He was very fond of me because I was wilder than my cousins. And I loved him for his many animals—horses, dogs, birds, and even a roe deer. When he was on his deathbed, I would draw horses and he would whinny. But he was particularly fond of bees. He had

150 colonies. He built a proper house for them with a sculpted gable and a red tiled roof. When I lay in the grass of his garden I used to hear the buzzing of bees. What I didn't realize then was that I was watching the flowers having sex. "Plants," my grandfather explained, "are anchored to the ground. They can't stroll across the meadows and hug each other and they can't have children without help. What they need is a love messenger—a bee."

It was a familiar and natural setup, rich in tradition, a fruitful synthesis between nature and culture; my grandfather even designed the labels on the cans. But his concept was too romantic; the competition worked more effectively and had fruit delivered by rail. The business suffered, and eventually the estate had to be sold. All that remained for my grandfather were the bees. They still brought him nectar from his lost lands where the apples still grew. The age of the "economic miracle" had arrived; trees were felled on the sunny slopes above the town to make way for sprawling residences. My grandfather despised the residents because despite having all the trimmings of nouveau riche luxury, they dined in the kitchen. He found accommodation in the summer residence of rich relatives, the Villa Eden. The gilt V of "villa" on its gate had long since dropped off and had never been replaced. When we children dared to approach the small bee house in the overgrown garden and listened to the excited buzzing in the summer heat, we found it a magical place, fascinating but also scary. In the midst of all this stood an old man with a straw hat and no protection. The bees didn't harm him; it was as if they knew him.

But even in this idyllic setting, beekeeping relied on a few tricks. It used to annoy me that my grandfather fed the bees a cheap sugar solution as a substitute for honey, rather like giving cheap beads to Native Americans in exchange for gold.

Now, globally, bees are doing really badly; they seem to have reached the ends of their powers. Alarmed by these reports, I, as grandson of a beekeeper, set off on a trip around the world to seek a solution to the mystery. Even the subsequent generation has become involved with the destiny of bees. My daughter and her husband, Boris Baer, work at the University of Western Australia researching genetics and the immune system of bees in the hope of breaking the vicious circle that is threatening to wipe bees out. The bees don't frighten my two small grandchildren, Andrin and Lucien, who often visit the bee labs looking like mini astronauts in their oversized bee suits. The younger one still takes his cuddly penguin with him beneath his armor—just for safety's sake. He particularly likes the smoker. Both of them often help with marking the newly hatched drones with different colors to signify particular birthdays. It's not that easy to make a tidy dot with a small paintbrush on the backs of drones. Andrin has found an imaginative solution to counter *Varroa destructors*, in his hand-drawn comic strips, the bees wear knights' armor and spikes protect the hives.

A narrative link exists in our family history from my grandfather, who was a beekeeper, via me to my grandchildren. I first got to know about bees as a child and now I talk about them with my grandchildren; they are the future. The movie is dedicated to all grandchildren. A member of the Club of Rome predicted that in twenty to thirty years' time, the young will start a revolution to reclaim the environment. I only hope that we can manage it quicker and in a peaceful fashion.

I first got to know about bees as a child and now I talk about them with my grandchildren; they are the future.

Initially, the extent to which the personal relationship should be visible in the movie wasn't clear: Can we show it or is it too intimate? In the end we decided yes, we could, because *More Than Honey* shouldn't be just a documentary. The personal relationship gave me more freedom to be subjective.

I never wanted to make documentaries, but in the 1960s, when I was starting out, you could forget making feature films because of the lack of sponsorship. This is why my first movie after graduating from film school was a documentary about the cavalry. It is loosely related to *More Than Honey* in that my grandfather was lucky enough to be assigned to the cavalry, the only chance for a 19-year-old to have something to do with horses. The movie was about whether horses liked being part of the military system as much as those who proudly rode them. The movie was banned but still won a number of prizes. Previously, while at film school, I had made a documentary about a prison that was banned because I had blanked out the faces of all the inmates while leaving the faces of the wardens and governor visible. They found that highly offensive and felt compromised. These conflicts strengthened my resolve to change to feature films. In the meantime, film sponsorship had recognized the value of fiction, so I remade the prison documentary as a movie (*Fluchtgefahr, Danger of Escape,* 1976), basing it on the diaries I had written while working incognito as a prison warden in preparation for the documentary. This wasn't so easy to ban, as it was "fiction."

Actors are at the heart of my work. It is almost an erotic pleasure when directing to witness how ideas suddenly become three-dimensional and begin to come to life. An individual playing with reality was one of the topics covered by a movie script about a con man, and my mind had been occupied with thoughts about individualism for years. But in all these egocentric excesses, which well matched the times of the artificially inflated economy, I could find no way of bringing the hero down to earth without destroying him.

The bees' swarm intelligence was a real release for me. As a documentary director, I had to closely watch my leading actors, the bees (and the human protagonists too), only being able to give a few instructions. This made life difficult, but it was an exciting challenge to try something completely new. Close scrutiny, necessary for a documentary, had already been an important aspect of my feature films and I always researched them thoroughly. The dramaturgical experience gained from feature films was a great help for my first full-length documentary. A documentary is still about telling stories that form an entity when linked together. I pinned up cards displaying all the individual scenes in order to find possible sequences, and especially to experiment with turning points. Afterward I made a sketch a meter and a half (almost five feet) long, rather like a kind of musical score, so that I could see at a glance the continuity of the movie.

Work on *More Than Honey* took five years; one year for research and project development, one year to organize the finances, two years' filming on four continents, and one year for editing. Despite my family background in beekeeping,

I still had to familiarize myself with the work. I set off on a trip around the world to get to know everything I needed to know. Using a small video camera, I recorded conversations and kept a diary. Initially I made the film in my head. This was important for the planning as the filming was dependent on dates precisely scheduled by nature and spread across four continents. Particularly in spring, there were many potentially critical overlaps: apple trees begin to blossom in the Shanxi province around the middle of April, then the pollen has to be harvested, but at that time we should be in Arizona because the desert heat is still bearable then. To film the hand-pollination, we would need to be in the Liaoning province around the middle of May, but then we would risk missing the foulbrood inspections in the Bernese Oberland.

I began researching in Australia with my daughter and son-in-law. Through the renowned Centre for Integrative Bee Research (CIBER) I made contact with researchers and beekeepers from other parts of the world. My research trip turned into a relay race as I was passed from one beekeeper to the next. My grandfather and my bee researcher in the family opened doors throughout the world.

In the USA I started with the scientific beekeeper Randy Oliver in California, a practical person who has published various articles based on his work and research and with whom I did a kind of bee apprenticeship. I lived and worked with him for three weeks in Grass Valley in the foothills of the Sierra Nevada. When we set off in his old truck to move the hives we took his mother with us. We dropped her off en route at Lake Tahoe for two days at the casino. The liveried doormen raised their eyebrows as our buzzing truck drew up at the entrance. Randy put me in contact with many other beekeepers, queen bee breeders, bee brokers, and bee dealers—and also with John Miller.

The first thing I saw on John Miller's honey farm was splitting, early in the morning and on an empty stomach. I could almost feel the brutality of this mechanical process; it was like being in a slaughterhouse, the floor covered in dead bees, huge swarms hanging on all the trees trying to escape, and clamor everywhere. If I hadn't had a net in front of my nose I would have been breathing in bees.

Miller had to leave early, and waited for me on a highway bridge with his pickup. He was making a call as I arrived and handed me a sandwich so that I had something to eat while he talked on the phone. Then we sat down in a café and the first thing he said was: "What we're doing here really is cruelty to animals."

So even at the onset he was tempting me onto thin ice with his openness, something that also comes across in the movie. He went on to explain that "I even want the bees to be stressed because then they are more likely to accept the new queen."

His humorous openness and honesty fascinated me; he, with his industrialized conveyor belt beekeeping, is one of the reasons why bees are in such a predicament. But he is also able to analyze and to assess critically, although each time in a typically American fashion he always finds an optimistic solution or an excuse. As a leaving present he gave me a photograph of my grandfather that he had had framed and to which he had added an inscription: "A shared experience." Through both our beekeeping grandfathers we had almost become friends, even when he asked me whether he had been cast in the role of Satan after having watched the trailer online. He is a Mormon and a marathon runner, and says that he only reads the *Wall Street Journal* and occasionally the *New York Times* to find out what the enemy is thinking. His son used to work for Apple but was lured into joining his business. Miller Honey Farms—California, Idaho, North Dakota is a dynasty. Miller has a brother in Idaho who is also a large-scale beekeeper, but they don't talk to each other. There is even a book about Miller, *The Beekeeper's Lament.* And there is a book about his whole family and his grandfather, who developed migratory beekeeping using the railways, *Sweet Journey: Biography of Nephi E. Miller, Father of Migratory Beekeeping.*

Nowadays they transport the bees across the continent in huge trucks and try to stop as little as possible. The entrance holes remain open throughout the journey with nylon nets spread over the cargo to prevent the bees from flying away. When, despite the airflow, it becomes too hot, they stop at a car wash and spray the load with cold water. The logbooks have to be adjusted,

A member of the
Club of Rome predicted
that in twenty to thirty
years' time, the young will
start a revolution to
reclaim the environment.

because legally it wouldn't be possible to travel the 2,700 kilometers (1,678 miles) in two days and a night with all the mandatory breaks.

During the journey we continued to film, changing from one truck to another and driving ahead in the escort car to film impressive landscapes, and at one stage even transferring to a helicopter. Our work was increasing the stress loads of the bees. When Miller rang us late at night to ask about progress he was furious: "You're endangering my bees and my drivers!" He doesn't love just money but also his bees. But he's driven by money, and his bees pay the price.

Prior to filming in the almond plantations, we had secretly hoped that we would be able to film one of the pesticide vehicles, but in the end it was difficult to keep them out of the frame. They were constantly audible, making it difficult to find any sort of peace for interviews and filming because they were simply always out and about. And then suddenly they would drive straight through our set. Miller becomes very angry when the bees are sprayed but here too he can find a way out: "They do what they have to do. It's a pact with the devil." The pollination industry, of course, also has its history, based on how humans selected bees in the first place. Initially we wanted to open the film with honey hunters in Nepal to show the oldest form of confrontation between humans and bees. The hunters abseil, with no protection at all, down cliff faces, poking at the dripping honeycombs until they drop. There are also Spanish cave paintings depicting this procedure. The bees naturally try to defend themselves, and in some cases the hunters get stung in their eyes and become blind.

Humans have learned to minimize such dangers. The first scene in the film, after the birth of the queen, now shows the insects domesticated by humans as farmed animals, an almost mythological act. Instead of an apple, though, Adam picks the bees that create the apple from a tree. The bees were actually the only creatures to have worked in Paradise.

The mountain valley in these clips is one of my favorite landscapes. During the Ice Age, there was a huge glacier here; Bronze Age people lived on its extremities, and you can still find traces of their dwellings on the alpine meadows. Then someone with a ladder and a box wanders into the picture and proceeds to capture something. Trudging home with his booty, he places it in something resembling a doll's house and latches the little door.

That is the beginning of the cooperation between bees and humans—domestication in the sense of bringing something home. It was the start of a difficult year for Fred Jaggi. Eventually he was forced to gas all his beautiful, pure-bred black bees, the strain of bees that he had cared for since his grandfather's time, which is precisely the reason that they were especially susceptible to diseases—there had been no fresh blood for such a long time. After gassed bees are burned, the empty cases have to be disinfected. Originally we wanted to film this in a nuclear power plant where the Swiss bee institute Schweizerischen Bieneninstitut was holding disinfection trials with gamma rays, but we couldn't get permission to film. Normally disinfection is carried out in the ovens of the nearest bakery, but the bakery was so ugly, a metallic barrack-like building selling enormous amounts of frozen croissants from Poland to tourists from the nearby hotels, that I didn't feel like using it for this scene, particularly as the burning of the beehives was so emotional. The fire burned for three days in the pit that Jaggi had prepared.

Jaggi was very depressed and even unsure about whether to carry on. Then I gave him three colonies, all from Swiss strains. From all his experiences, however, he had become uncertain as to whether the thoroughbred ideal was really such a good idea and so he requested Carniolan queens. He has also switched to a broader gene pool and now has twenty-three healthy colonies, including a number of hybrids.

I discovered the Singer family, who breed Carniolan queens, online. They are very energetic and enterprising, and have an impressive online presence. When I first encountered them, Heidrun Singer was still married to an operetta singer from the Vienna Volksoper and was using her married name of Luftensteiner-Singer. In reality she is married to the *carnica* and the family brand is called Carnica-Singer. As she says, "I'm only here for the bees."

Heidrun Singer is the chairperson of the Austrian apiarists, has already been presented to the federal presidents, and has hives on the roofs of the

Vienna Staatsoper (the state opera), the Burgtheater, and the golden dome of the Vienna Secession.

Once a year she organizes a race between a mountain biker and a bee on a mountain near Mariazell. The bee is marked and starts at the same time as the mountain biker at the top of the mountain. The bee usually wins. At the mountaintop restaurant there is a stuffed bear that Singer's father shot in the former Yugoslavia.

I found it particularly appropriate that the breeding of queen bees was a business enterprise run by women: the elderly mother, Liane Singer; her daughter Heidrun; and, although now only occasionally as she is in school, Liane's younger daughter who is in charge of the royal jelly side of the business. One of Heidrun's brothers is a large-scale beekeeper and the other a pilot, which also seems appropriate. The patriarch of the family is 84-year-old Wolfgang Singer, who was honored for services to agriculture. In earlier days he was the official bee inspector of the region and now often sits beneath a tree with his field glasses, checking that everything is being done correctly.

I found it particularly appropriate that the breeding of queen bees was a business enterprise run by women.

We filmed Heidrun delivering the bees to the post office in the neighboring village to be sent on to customers. But as we tried to film their arrival at the central sorting station in Vienna, where they were to be forwarded to addresses around the world, we were informed that the post office didn't handle living animals. We were able to prove the opposite by producing the original cases with stamped labels and were then allowed to film.

This demonstrates the absurdity of the bee smuggling story at the Swiss border: Bees can be sent and delivered throughout the world but if they are smuggled across borders they have to be destroyed in the name of disease limitation.

How could I portray a future without bees in the film? Film sequences cannot be filmed in the future, the camera only records the present. Together with director Ulrike Koch, who filmed *The Saltmen of Tibet* and supervised the casting of extras in Bertolucci's *Little Buddha*, I traveled through four Chinese provinces in the quest for footage of hand-pollination of apple plantations. It was a memorable journey that deserved a movie of its own.

It was not particularly easy to get any precise information in China. Somehow the whole subject seemed to cause some discomfort, especially as foreigners were asking the questions. An official bee representative in Sichuan, who called himself Bee-king in his emails, gave us the best tips in the end.

Of all the people we met, we eventually focused on Zhang Zao from the northern province of Liaoning, because she was a human bee. As there are only five days of apple blossom in the north, there is not enough time to harvest the pollen, process it, and get it to the blossom, so she travels two thousand kilometers (1,243 miles) to the south where the apples blossom a month earlier. She gathers the pollen there and travels back north in time to pollinate the apple blossoms. Her story illustrates the immense efforts and costs that would arise were bees to vanish.

On the freeway to Liaoning we saw people sweeping the road with dustpans and brushes. China has a very different approach to manpower.

Toward the end of filming, Zhang Zao became more and more reticent. She wanted to know why I wanted so many details and why she should betray the trade secrets of the People's Republic of China. She stopped speaking to us and later refused to cooperate at all. Long negotiations deep into the night were unsuccessful. A generous sum to cover expenses swayed her resolve, but finally our offer became seen as further proof that something was not quite right and she demanded to see our official permit. We didn't have one. On the advice of our Chinese production partner we had entered China as tourists. "Tomorrow morning I will go to the authorities and make a report," was her response. This was very dangerous for all of us but especially for our Chinese crew who had made filming possible. The Chinese camera assistant hid the disk with our recordings in his underpants and we made a bolt for it.

But nature had other surprises for us—the "killer bees." Friends teased me at the beginning about whether I wasn't just a bit frightened at the prospect of "killer bees." I laughed at them. For me they were an invention of scriptwriters

to fuel the fears of invasion during the Cold War. At my first Apimondia international bee congress, in Montpellier, I learned that "killer bees" make honey.
Professor Gonçalves from the Universidade de São Paulo—where Kerr inadver-
tently created "killer bees"—gave me the details.

Then I began to search online for people in the southern states of the USA
who work with them. I found offers from so-called bee wranglers advertising
pest controls with Hard Rock logos or bee-riding rodeo beekeepers. It seems
the whole business of pest control is still the realm of "killer bee" B-movies.

Randy Oliver, my scientific beekeeper in Grass Valley, also had a friend in
Arizona, Fred Terry. He was the ideal US antithesis to John Miller.

"Killer bees," or more precisely Africanized honeybees, react very aggressively to black, which is why beekeepers should always choose white
facemasks. The trouble was that I couldn't film Terry's face through the white
veil, so I asked him to wear a black one. Quick movements, dark colors, and
wool are the most provocative things of all. This proved to be the case with
our microphones' wind protectors; dark and fleecy, the bees laid into them as
if they were bears.

Our first day of filming in Arizona was of the honey harvest. Terry drove
his clapped-out old truck through a gate into a pasture and donned his protective gear, although there were no signs of bees. Using twigs and leaves, he then
fired up his huge smoker, which was three times as big as the European ones
and needed two hands to hold it. That was the job of Fred's pretty girlfriend,
Marie, who could have been Joan Baez's sister. There followed the command
"APF"—always pee first. You dread to think of the alternatives. Then they drove
in their suits to the site of the beehives. On getting out of the truck you
could already hear that something was different, the tone was much higher
and more aggressive than normal flight sounds. When Fred opened the cases
things really livened up.

Beekeepers wear long trousers and a long shirt under their suits as bees can
sting through outer layers. Every sting relays olfactory information: "Watch out!
Danger!" When a bee stings, all the other bees in the vicinity come to help out. I
love my kid leather gloves, but here in the desert you wear clumsy, thick rubber
gloves. For safety, you can also tape up the cuffs of gloves and the tops of boots.

We thought about cutting a hole for a lens in the cameraman's facemask, but this proved to be too much of an inconvenience and somewhat danger-ous, so we decided against it. When the cameraman tried to look through the lens—it was impossible to view the monitor in the bright light—the viewfinder pressed the net to his nose and he promptly got five stings on the nose that became horribly swollen. We then tried to find a mask with an elongated nose, rather like a harlequin mask, so that the net no longer touched his nose. This was not an easy task in Tucson.

When we drove off after gathering the honey, we were followed by an angry cloud of bees for 1.5 kilometers (almost one mile), which is why it probably wasn't such a bad idea to have the beehives locked in an enclosure far away from civilization. And just when you take off your hood in front of a Mexican bar and are looking forward to a meal, a bee creeps out of a fold in your cloth-ing and stings you. Fortunately, bees somehow seem to be in our family blood. My team was always somewhat jealous that when I was stung it didn't swell; they said that I had inherited resistance and that we should make a serum from my blood for people with allergies.

We would really have liked to film bears hunting honey but our budget wouldn't allow it. However, we at least wanted to film the upturned, empty hives after a bear raid—just the sort of scenes that you would imagine if you weren't a documentary maker. I secretly wanted the "killer bees," which Fred Terry had retrieved from a roof and rehoused at the beginning of the scene, to return to the wild in the end, a kind of reversal of the opening of the film with Jaggi and the bees that he domesticated. Then, as if they had read the script, the bees simply vanished all on their own. Fred Terry was astonished.

All that remained for us was to film the cactus scene, where the bee swarm had settled before they moved on to a cliff face in a canyon and freedom. The bees played along brilliantly.

Working with bees as a director was new to me, but it fulfilled an old dream. Originally I wanted to take German studies and zoology at university, but this combination was not considered an option by the university as the main lectures for both subjects ran concurrently between eleven and twelve o'clock. And then my daughter, Barbara, took over this part of my life. Even as a child she was always bringing creatures home, so it seemed logical that she would become either a vet or a biologist. While working on her doctoral thesis on intestinal parasites of bumblebees at the ETH Zurich, she met Boris Baer, who was writing his thesis on the genetic advantages of promiscuity in bumblebees. They have now both been researching bees for eight years at the University of Western Australia in Perth.

Recently the *Varroa* mite has infiltrated New Zealand with devastating results. Beekeepers didn't know how to deal with this plague and could only look on in despair as their colonies died. They only slowly recovered from the shock. In Australia, of course, everyone was anxious about what was in store for their bees. Can you prepare for something that doesn't yet exist?

Crossbreeding domesticated bees with wild honeybees to strengthen the gene pool, as practiced by Boris Baer and Barbara's team, is almost like taking a step back in time to the experiments Professor Kerr made in Brazil that got out of hand and resulted in "killer bees." To preclude this risk, the Australian test bees were brought to an uninhabited island. Should Frankenbees result from these experiments, no one would be endangered, as the mainland is too far for them to reach.

Unfortunately, all the laboratory equipment for analyzing the bees' DNA, which cost millions, was less photogenic than the trials in the field but was, of course, fundamental to the experiment as a whole. The studies were so interesting that the University of Copenhagen even sent human sperm to the Australian team to undergo the same sort of analysis as the drones.

As I visited the lonely island of the test bees for the first time during my research, one and a half years prior to filming, it dawned on me that here an almost utopian Noah's Ark was coming into being; it was a spot where the last healthy bees survive, whereas everywhere else in the world they are struggling. The situation was even more eerie because the island is a restricted military zone with underground chambers for submarines, all of which was out of bounds for cameras, of course.

The initial research journey resulted in many ideas and encounters with additional protagonists who were dear to me, especially in the industrial sector; there was a bee agent in the USA and a beekeeper in Australia who dealt in package bees—bees posted in cardboard tubes and destroyed after pollination to save costs. These people, however, were not too keen on being filmed. And then there were the greenhouses in which we did film; viewers just can't imagine how strawberries grow there and that they have to be pollinated there. It was, in a way, a repeat of the Miller story; the protagonist was not completely convinced about the problems of his greenhouse beekeeping methods. This was why we decided to cut it from the film and keep it for the extras section on the DVD. There was the idea of following an almond from the plantation to its resting place in the finished product of a cookie—all this madness for a couple of cookies. And we discussed the extent to which the world economy should play a more concrete role, but I believe that it was right to omit all of that. The topics and concepts are lurking behind everything you see. Then again, expanding to include the pesticide industry and the political struggles and disputes would have made the whole thing more journalistic. Eventually we dispensed with these thoughts; the movie should trigger these discussions and not portray them. You can witness the effects of pesticides close up in the film, so there is no need to name the agrochemical companies that have rejected any form of contact with me for five years. Their refusal is more effective than an admission of guilt.

The human protagonists should all stand for themselves and be taken seriously, but they also document a specific stance on bees. The differences between the central characters were important both to tell the story and to encourage debate. But the central question was: Who are the protagonists and who are the antagonists of the movie? Are bees or humans in the starring role? Gradually, the bees became the stars, and the movie begins and ends with bees.

But I couldn't ask the bees questions, I could only observe them and imagine what they felt and thought. Do they even think at all? The bee researcher Professor Randolf Menzel was an invaluable contact in that respect. He knows bees inside out and is fascinated by the fact that these creatures with their minuscule brains can actually make decisions, something he can prove scientifically.

Our discussions, both at the institute at the Freie Universität, Berlin, and on the grass at his trial site at Klein Lüben, often revolved around the role of the individual in a swarm and the question of the extent to which we as humans are allowed to try to "think" like bees. This question bothers him as a scientist who has to objectively analyze solid facts more than it does me as a film director who is free to let emotions influence matters. His answer was explicit: As humans, the only instruments that we have at our disposal to research the bees' brains are our own brains, but we have to be very conscious of this when we are trying to "think" as bees.

So, how do you escape accusations of having anthropocentric views, known by the rather disparaging term of "pathetic fallacy"? By saying that bees are simply robots and giving them only the reflexes of stimulus and response, the mechanical reactions to external influences? There is even the extreme, scientific-materialist approach that declares that you cannot love the subject of your research because emotions tend to distort objectivity. Luckily, Menzel has a completely different opinion and says very clearly that he loves bees, even to the point of exclaiming that "the colony as a whole also has feelings!"

It is undisputed that we humans have feelings and empathy, and as Aristotle knew, empathy is one of the basic requirements of dramatic theory. There is no such thing as an objective film; even a documentary is an expression of a personal position. As soon as I place the camera somewhere I am making a personal statement.

When I started at film school there were no videos to practice on; from our first year we had to film with a 35 MM blimp camera weighing seventy kilograms (154 pounds). "Think carefully about where you place it," laughed the teacher. That was an experience with far-reaching effects, finding a particular position, first mentally and then literally, and I learned to let the camera do the moving.

In my movies I accuse people of overcontrolling and of intervening in a manipulative fashion—and in order to formulate this I have to practice it myself. That is part of the issue in the first place. We are all involved in making the world what it is, there is no single culprit that we can all point at—we are all responsible. It wasn't easy to formulate this without being too rosy or too smug. When I say that the bees were the protagonists, I accept potential accusations of anthropocentrism, but I didn't want to make a cold, scientific movie. We were making a movie for the public, a cinema movie; it needed an identifiable subject so that I could tell my story. Letting my bees talk would have been going too far, but I did allow myself to playfully ask, in the "killer bees" scene, whether the bees were finally fighting back. I am pleased about their rebellion, that they are no longer making life easy for us. But how do I persuade my viewers of this? How can I get them to love bees too? Die Biene Maja (better known in English-speaking countries as Maya the Bee) has a good reputation for sure, but many people are still afraid of insects. As they are leading characters in our movie, we have to film them not just as a blurred mass but as individuals. They needed to be given a face.

Their world is usually hidden to the human eye. The macro scenes had to precisely blend with the documentary scenes; the viewer should be able to observe exactly what happens to the bees throughout all the various strands of the storyline, as the bees experience it. This is action from the perspective of bees. For instance, when the bees are traveling across the USA in huge trucks, we should be able to see and experience what it is like in those shaking cases, where new bees are hatched during the journey, and to see the parasites that clamber about on their heads. But we should also be able to see the queen in its narrow transport cage being placed in a yellow envelope and then stamped at the post office.

With a team of specialists, we constructed a macro studio in the "Alten Brotfabrik" on the outskirts of Vienna, where we could accommodate our own fifteen colonies of various strains on the large overgrown grounds of the Brotfabrik. During our two-month stay there, we only filmed macro scenes. In addition, the macro team visited various other areas in Austria and also Arizona—we couldn't film the "killer bees" in Vienna with a polystyrene cactus, we needed a real desert and real Africanized bees.

For the "human" part of the film I needed a "hunter" who could quickly react to the protagonists while linking the people with the landscapes and the bees. Enter Jörg Jeshel. For the macro clips, we needed a cameraman with a totally different character, a "gatherer," someone with patience and who would take pleasure in immersing himself in the world of bees, a place of creativity and constant activity. This was carried out by Attila Boa and a much larger team than Jeshel needed. Filming for the human aspects of the film required five people; for the macro slots, numerous people were needed at various times to film one single bee—the cameraman, two camera assistants, a script supervisor, a lighting technician, a technician for tracking shots, drone pilots to operate the mini helicopters with remote controls, carpenters for the props, a recording director, the bee carer, and me. We needed as great a variety of high-speed cameras and endoscopic lenses as they do in operating rooms.

In the process, a huge number of technical problems had to be solved that sometimes also raised fundamental questions. From the very beginning it was clear that we couldn't film the bees at their usual speeds. We experimented for a long time to find the speed that was most appropriate. In the tests we discovered that the bees at 70 frames per second (FPS) move roughly as quickly as humans. Viewers shouldn't have the feeling that everything is happening in slow motion. It should be taken for granted that you can observe the bees naturally and that at 70 FPS you can see exactly what they are doing. If you film at 24 FPS, the small bee's flurry of movements—the tongue, the feelers, and the wings—all become impossible to appreciate in detail.

All flying bees were filmed at 300 FPS as it seemed to us that the wing movements have a more natural appearance with a certain amount of blurriness; the effect is more poetic than clearly defined wing movements. Bee wings move at a rate of 250 beats per second. For us, 24 FPS results in smoothly flowing images, but this occurs for bees only at 280 FPS because every single compound eye registers a different image, the next compound eye again a different one and so on. Only at 280 FPS do bees experience a smooth flow of motion. I discovered all this later, after our experiments with the wings. So, our 300 FPS was also the speed that bees see as flowing movement and no longer as a stroboscopic effect.

For the macro filming in particular, I was dependent on the experts. Our "bee whisperer," Peter Hopfgartner, was one of those experts and was also called upon for filming in the USA. He is a beekeeper himself, is on his second course of studies in philosophy, and knows the language of bees, that is, he knows what they are about to do but cannot give them orders. This is why we filmed in April/May: there is plenty going on in the bee world at that time.

We had a long list of topics that we wanted to cover: nectar deliveries at the hive, depositing and storage of pollen, the waggle dance, the building of the honeycomb, the birth of the queen, and so on. Our bee whisperer inspected the various colonies to find out where what we needed was actually happening. In the meantime, we had prepared everything in the studio—an empty fabricated honeycomb, lights, cameras, and all the technical requisites. Then Peter Hopfgartner brought in the frame and we could only hope that it would still be happening, or would happen again so that we could get our shot. In both instances we didn't have much time as the bright spotlights created an environment that was markedly different from the dark hive and our technology proved a major distraction to the animals. First of all we looked at the honeycomb with our bare eyes, because when looking through the macro optic you miss a lot of what is happening in the margins. Then we decided which action we would concentrate on. Once this had been decided, I followed everything on the big monitor and could tell the cameraman what was interesting me the most. At the same time, I let the script supervisor know what should be noted down and by which time code (FPS), and then made my own notes for the cutting room.

All this requires an endless amount of material. It takes a lot of time to find the right bee with only the tiny detail seen in the endoscope or one of the other macro lenses. If you only switch on the camera after you find the right section, you are too late. We could manipulate some things a little, but only to a very limited extent. Sometimes totally unexpected things happened while we were filming, and we had to act quickly to accommodate the changes. On some days, we managed only a few seconds of usable film; on others, the bees were kind to us. We worked for a whole week for the mating scene, which lasts all of thirty-six seconds. In the end, we had 105 hours of macro material, twenty-five minutes of which appear in the final cut.

Heat generation was a particular problem. High-speed cameras need much more light; the more frames per second that have to be exposed, the brighter the lighting has to be, and the brighter the lighting, the hotter the lamps. This, of course, was a challenge because we were working with wax and we had the bees' welfare to consider. Sometimes we filmed outside using mirrors, as the sun was brighter than our spotlights and we could use the wind to cool down the equipment.

Considering the effort required, it wasn't surprising that in the end we needed another year to make use of the brief window of April/May to get the necessary recordings of the bees in all their diversity on film. Filming a bee gathering pollen, something that we thought was going to be the easiest thing in the world, turned out to be very challenging. When we wanted shots of a certain blossom we had to let the cameras run and then either wait and wait and wait—often in vain—until a bee eventually arrived, or try to encourage the bee onto the blossom—also mostly to no avail as their flower-fidelity was programmed for other blossom types. If you took a bee from the hive entrance and placed it on a blossom, first the flight and landing on the blossom would be missing, and second, most of the bees flew straight back to the hive as they were guard bees and had nothing to do with foraging, or it was a nurse bee that also didn't feel like collecting pollen because it already had another job. It would be like employing a dressmaker in a bakery.

Getting the entire richness of this division of labor on film was a Sisyphean task but also a fascinating experience. Take, for instance, the nuptial flight with the camera flying at the same height as the queen, or the workers releasing the queen-to-be from its brood cell with the camera circling the event capturing the presence of the queen's colony. This should give the viewer the impression of being on the scene, and almost of being required to lend a hand. We had to predict exactly when the queen would hatch, otherwise we would never have had all the cameras and technology in place for the birth.

For the aerial shots we used mini helicopters, motorized drones with small cameras. Many advisors told me that this was far too complex and that instead we should create a 3D bee that could do everything and didn't actually sting, but there was no way I was going to agree to that. All the bees that you see in the film are real. We worked a lot with smells so that we could communicate with them and persuade them in their own language, but the most effective trick was patience.

We could certainly plan a lot, but with a shooting period of almost one hundred shooting days over two years not everything was practicable and we had to be ready for anything and everything. The editing phase was crucial to creating a ninety-minute story from 250 hours of material. Most of the changes to the screenplay happened during editing. The fresh eyes of our film editor, Anne Fabini, were important as she had never seen the real thing, only the recorded images from which the story had to be told. Anne Fabini originally wanted to be a midwife and deliver babies; instead she became a crucial player in the delivery of the movie.

In the movie there were some things that we didn't want to state directly, we wanted the viewers to have the opportunity to draw their own conclusions from all that they had seen. But in order to play the game, all the cards have to be on the table. Thinking ahead about the amount of information and the playing options for the viewers was our challenge.

The statement that humankind cannot survive without bees may well have only been attributed to Einstein, but another comment that does come from him fascinated me more and more given my preoccupation with bees: "We cannot solve our problems with the same level of thinking that created them." This is why I am trying to optimistically prove that the quote attributed to Einstein is wrong and that we are more likely to die and the bees to survive. But *More Than Honey* is not a catastrophe movie. After studying the case history, we look at the efforts to find a cure, the prospects of finding one, and the unexpected solutions that nature could provide.

Over the thousands of years of the relationship between humankind and the bees there have been increasing numbers of conflicts between civilization and nature. This raises fundamental questions: Is humankind a part of nature? Or do we just want stand above it and subdue it? Could there not be some form of fruitful symbiosis between all parties—the bees, the beekeepers, the plants, the farmers, the dealers, and the eaters—a kind of Allstar Jazz Orchestra with various soloists appreciating one another in order to play music together? A utopian model of swarm intelligence.

Markus Imhoof, Berlin, September 2012

Notes

PREFACE

1. See "Milbe verbreitet tödliches Virus unter Bienen," SpiegelOnline, June 8, 2012. www.spiegel.de/wissenschaft/natur/bienensterben-milbe-verbreit-et-toedliches-virus-unter-bienen-a-837744.html.

2. United Nations Environment Programme, UNEP Emerging Issues: Global Honey Bee Colony Disorder and Other Threats to Insect Pollinators, 2010, p. 7. www.unep.org/dewa/Portals/67/pdf/Global_Bee_Colony_Disorder_and_Threats_insect_pollinators.pdf.

3. "Bienensterben wird zum globalen Problem," SpiegelOnline, March 10, 2011. www.spiegel.de/wissenschaft/natur/uno-bericht-bienensterben-wird-zum-globalen-problem-a-750139.html.

4. Richard Friebe, "Volk der Bienen, quo vadis?," Frankfurter Allgemeine, April 6, 2011. www.faz.net/aktuell/wissen/natur/bienensterben-volk-der-bienen-quo-vadis-1622343.html.

CHAPTER 1

1. "Bee Mites Suppress Bee Immunity, Open Door for Viruses and Bacteria," ScienceDaily.com, May 18, 2005. www.sciencedaily.com/releases/2005/05/050517110843.htm.

2. "Parasiten-Fliege könnte Bienensterben auslösen," SpiegelOnline, January 4, 2012. www.spiegel.de/wissenschaft/natur/insekten-parasiten-fliege-koennte-bienensterben-ausloesen-a-807031.html.

3. Julius Kün-Instituts, "Analyse des Julius Kühn-Instituts zu Bienenschäden durch Clothianidin." Press release. June 10, 2008.

4. Jürgen Langenbach, "Neubewertung des Einsatzes von Neonicotinoiden bei blühenden Pflanzen," *Die Presse*, March 29, 2012.
5. Atlant Bieri, "Gifte in Schrebergärten und auf Feldern bedrohen Bienen-völker," *Tages-Anzeiger*, April 22, 2012.
6. Bernd Welz, "Die Wahrheit über das Bienensterben," *natur+kosmos* 7 (2009).

CHAPTER 2

1. Rudolf Steiner, "Über das Wesen der Biene." In *Mensch und Welt. Das Wirken des Geistes in der Natur. Ueber das Wesen der Bienen. Vortraege fuer die Arbeiter am Goetheanumbau*, Rudolf Steiner [Fifteen lectures presented to workers at the Goetheanumbau in Donach from 8th October to 22nd December 1923, Dornach 1988].
2. M. Lehnherr and H.-U. Thomas, "Natur- und Kulturgeschichte der Honig-biene." In *Der schweizerische Bienenvater*, ed. Verein deutschschweizerischer und rätoromanischer Bienenfreunde (volume 5, 18th edition, Winikon: Fachschriftenverl. VDRB, 2003), 40.

CHAPTER 3

1. B. Lehnherr and N. Duvoisin, "Biologie der Honigbiene." In *Der schweizeri-sche Bienenvater*, ed. Verein deutschschweizerischer und rätoromanischer Bienenfreunde (volume 2, 18th edition, Winikon: Fachschriftenverl. VDRB, 2003), 81.
2. cf. W. Jacobs and M. Renner, *Biologie und Ökologie der Insekten*, 2nd edition (Stuttgart: Fischer, 1988), 41.
3. "Auch bei Honigbienen gibt es Draufgänger und Angsthasen. Persönlich-keitsunterschiede zeigen sich im Verhalten und am Gehirn," *scinexx* 9 (March 2012). scinexx.de/wissen-aktuell-14537-2012-03-09.html.
4. Francis Heylighen, "The Global Superorganism: An evolutionary cyber-netic model of the emerging network society," *Social Evolution & History* 6 (2007): 57–117.

CHAPTER 4

1. Ruedi Ritter, "Königinnenzucht und Genetik der Honigbiene." *Der schweizerische Bienenvater*, ed. Verein deutschschweizerischer und rätoromanischer Bienenfreunde (volume 3, 18th edition, Winikon: Fach-schriftenverl. VDRB, 2003), 73.
2. Bruder Adam, *Auf der Suche nach den besten Bienenstämmen* [*In Search of the Best Strain of Bees*] (Oppenau, 1983 edition).
3. A. Matzke and S. Bogdanov, "Bienenprodukte und Apitherapie." In *Das schweizerische Bienenbuch*, ed. Verein deutschschweizerischer und rätoro-manischer Bienenfreunde (volume 4, Appenzell: Neuauflage [reprint], VDRB, 2012), 60.

CHAPTER 5

1. cf: Kerstin Hoppenhaus, "Die Biene und das Biest," *Die Zeit* 50 (2011).
2. Franco Zecchin, *Bees in Paris*. www.picturetank.com/___/series/b773991aac75a12c6b8dafb20e716a70/Bees_in_Paris.html.
3. Cited by Martin Rasper, *Vom Gärtnern in der Stadt* (Munich: oekom verlag, 2012), 61.

CHAPTER 6

1. Ulli Kulke, "Der Mann, der die Killerbienen züchtete," *Die Welt*, April 12, 2005. www.welt.de/print-welt/article644721/Der-Mann-der-die-Killerbi-enen-zuechtete.html.
2. Ibid.
3. Ibid.
4. S. Berg, J. Schmidt-Baily, and S. Fuchs, "Varroabekämpfung mit Drohnen-brutfangwaben. Ein biotechnisches Verfahren," *Imkerfreund* 5 (2000): 6.
5. R.A. Calderón et al., "Reproductive Biology of *Varroa* Destructor in Afri-canized Honey Bees (*Apis mellifera*)," *Experimental and Applied Acarology* 4 (2010): 286.

6. L. Mondragón, M. Spivak, and R. Vandame, "A Multifactorial Study of the Resistance of Honey Bees *Apis Mellifera* to the Mite *Varroa Destructor* over One Year in Mexico," *Apidologie* 36 (2005): 345-358.

7. Calderón et al., 287.

8. Calderón et al., 290.

9. Arbeitsgemeinschaft des Instituts für Bienenforschung e.V. (ed.), *Varroa unter* Kontrolle (Passau, 2nd edition, 2007), 5. Available at staff-www. uni-marburg.de/~ag-biene/files/varroa_unter_kontrolle.pdf.

CHAPTER 7

1. Rudolf L. Schreiber (Ed.), Tiere auf Wohnungssuche. Ratgeber für *mehr Natur am Haus* (Berlin: Deutscher Landwirtschaftsverlag, 1993), 249.

References

Arbeitsgemeinschaft des Instituts für Bienenforschung e.V. (Ed.). (2007). *Varroa unter Kontrolle*, 2nd edition. Passau. Available at staff-www.uni-marburg.de/~ag-biene/files/varroa_unter_kontrolle.pdf.

Berg, S., J. Schmidt-Baily, and S. Fuchs. (2000). "Varroabekämpfung mit Drohnenbrutfangwaben. Ein biotechnisches Verfahren." *Imkerfreund* 5.

Bieri, Atlant. (April 22, 2012). "Gifte in Schrebergärten und auf Feldern bedrohen Bienenvölker." *Tages-Anzeiger*.

Bruder Adam. (1983 edition). *Auf der Suche nach den besten Bienenstämmen.* Oppenau.

Calderón, R.A. et al. (2010). "Reproductive Biology of *Varroa* Destructor in Africanized Honey Bees (*Apis mellifera*)." *Experimental and Applied Acarology* 4.

Friebe, Richard. (April 6, 2011). "Volk der Bienen, quo vadis?" *Frankfurter Allgemein*. www.faz.net/aktuell/wissen/natur/bienensterben-volk-der-bienen-quo-vadis-1622343.html.

Heylighen, Francis. (2007). "The Global Superorganism: An evolutionary cybernetic model of the emerging network society." *Social Evolution & History* 6.

Hoppenhaus, Kerstin. (2011). "Die Biene und das Biest." *Die Zeit* 50.

Jacobs, W. and M. Renner. (1988). *Biologie und Ökologie der Insekten*, 2nd edition. Stuttgart: Fischer.

Julius Kün-Instituts (June 10, 2008). "Analyse des Julius Kühn-Instituts zu Bienenschäden durch Clothianidin." Press release.

Kulke, Ulli. (April 12, 2005). "Der Mann, der die Killerbienen züchtete," *Die Welt*. www.welt.de/print-welt/article644721/Der-Mann-der-die-Killerbienen-zuechtete.html.

Langenbach, Jürgen. (March 29, 2012). "Neubewertung des Einsatzes von Neonicotinoiden bei blühenden Pflanzen," *Die Presse*.

Mondragón, L., M. Spivak, and R. Vandame. (2005). "A Multifactorial Study of the Resistance of Honey Bees *Apis Mellifera* to the Mite *Varroa* Destructor over One Year in Mexico." *Apidologie* 36.

Rasper, Martin. (2012). *Vom Gärtnern in der Stadt.* Munich: oekom verlag.

Schreiber, Rudolf L. (Ed.). (1993). Tiere auf Wohnungssuche. Ratgeber für *mehr Natur am Haus.* Berlin: Deutscher Landwirtschaftsverlag.

ScienceDaily.com. (May 18, 2005). "Bee Mites Suppress Bee Immunity, Open Door for Viruses and Bacteria." www.sciencedaily.com/releases/2005/05/050517110843.htm.

scinexx.de. (March 2012). "Auch bei Honigbienen gibt es Draufgänger und Angsthasen. Persönlichkeitsunterschiede zeigen sich im Verhalten und am Gehirn." scinexx.de/wissen-aktuell-14537-2012-03-09.html.

SpiegelOnline. (March 10, 2011). "Bienensterben wird zum globalen Problem." www.spiegel.de/wissenschaft/natur/uno-bericht-bienensterben-wird-zum-globalen-problem-a-750139.html.

SpiegelOnline. (January 4, 2012). "Parasiten-Fliege könnte Bienensterben auslösen." www.spiegel.de/wissenschaft/natur/insekten-parasiten-fliege-koennte-bienensterben-ausloesen-a-807031.html.

SpiegelOnline. (June 8, 2012). "Milbe verbreitet tödliches Virus unter Bienen." www.spiegel.de/wissenschaft/natur/bienensterben-milbe-verbreitet-toedliches-virus-unter-bienen-a-837744.html.

Steiner, Rudolf. (1923/1988). *Mensch und Welt. Das Wirken des Geistes in der Natur. Ueber das Wesen der Bienen. Vortraege fuer die Arbeiter am Goetheanumbau.* Dornach.

United Nations Environment Programme. (2010). UNEP *Emerging Issues: Global honey bee colony disorder and other threats to insect pollinators.* www.unep.org/dewa/Portals/67/pdf/Global_Bee_Colony_Disorder_and_Threats_insect_pollinators.pdf.

Verein deutschschweizerischer und rätoromanischer Bienenfreunde. (Ed.). (2003). *Der schweizerische Bienenvater.* Volume 2, 18th edition. Winikon: Fachschriftenverl. VDRB.

Verein deutschschweizerischer und rätoromanischer Bienenfreunde. (Ed.). (2003). *Der schweizerische Bienenvater.* Volume 3, 18th edition. Winikon: Fachschriftenverl. VDRB.

Verein deutschschweizerischer und rätoromanischer Bienenfreunde. (Ed.). (2003). *Der schweizerische Bienenvater*. Volume 5, 18th edition. Winikon: Fachschriftenverl. VDRB.

Verein deutschschweizerischer und rätoromanischer Bienenfreunde. (Ed.). (2013). *Das schweizerische Bienenbuch*. Volume 4. Appenzell: Neuauflage [reprint], VDRB.

Welz, Bernd. (2009). "Die Wahrheit über das Bienensterben." *natur+kosmos*, 7.

Zecchin, Franco. (n.d.). Bees in Paris. www.picturetank.com/___/series/ b773991aac75a12c6b8dafb20e716a70/Bees_in_Paris.html.

Index

Adee, Richard, 5

AFB. *See* American foulbrood

Africanized honeybees (AHBs), 63, 95–100, 102–5, 107–9

agriculture, industrialization of, 92–93

agrochemicals, ix, 4–5, 8, 11–12, 85. *See also* neonicotinoids; *individual chemical names*

agrotoxins. *See* agrochemicals

AHBs. *See* Africanized honeybees

almonds, 2–4

American foulbrood (AFB), 28, 85, 115

anatomy, 32

Andrena hattorfiana (mining bees), 121

antibiotics, 8, 17, 28, 85

Apis cerana (Asian honeybee), 63–64, 80, 82–84, 100, 114, 123

Apis mellifera (European honeybee), 16, 23–24, 83–84, 123. *See also Apis mellifera*, subspecies of

Apis mellifera, subspecies of: *Apis mellifera adamii* (Cretan bee), 24; *Apis mellifera anatoliaca* (Turkish bee), 24; *Apis mellifera carnica* (Carniolan bee), 24–25, 71, 104;

Apis mellifera caucasica (Caucasian bee), 24, 77; *Apis mellifera cecropia* (Greek bee), 24; *Apis mellifera cypria* (Cypriot bee), 24; *Apis mellifera iberica* (Spanish bee), 24; *Apis mellifera intermissa* (Morrocan bee), 24; *Apis mellifera lamarckii* (Egyptian bee), 104; *Apis mellifera ligustica* (Italian bee), 24–25, 95–96, 104, 108; *Apis mellifera macedonica* (Balkan bee), 24; *Apis mellifera mellifera* (European dark bee), 22–23, 69, 104; *Apis mellifera scutellata* (African bee), 95–96

apitherapy, 88

Apocephalus borealis (fly species), 11

Armbruster, Ludwig, 69–70

Armstrong, Lance, 7

Asian honeybees. *See Apis cerana*

Avarguès-Weber, Aurore, 54

Baer, Boris, 114–17, 128, 140

Baer-Imhoof, Barbara, 139–40

Bayer (company), 11–12, 15

bee brokers, 5

beehives, 6, 34, 56, 75

beekeepers and beekeeping, 5, 17–18,
22, 34–36, 59–60, 73–75, 109–10. *See
also* migratory beekeeping; urban
beekeeping

Beekeepers for Obama, 5

bee products. *See* apitherapy; honey;
propolis; royal jelly; wax

bees. *See specific subjects*

Bengsch, Eberhard, 71

Benjamin, Alison, 16

biogas industry, 93

Boa, Attila, 144

Bonmartin, Jean M., 14

Brazil, 95–100, 108

breeding, 19, 23–24, 34–36, 60–71. *See also*
crossbreeding; mating and reproduction

Brother Adam (Karl Kehrle), 69–71

Bt corn (GMO), 17

Buckfast bees, 70–71

bumblebees, 40, 81, 122–23

California, 1–6

cantharophily (pollination by beetles), 40

Carnica-Singer queen bees, 68, 135–36

Centre for Integrative Bee Research (CIBER),
114–17, 131

Cheeseman, John, 49

China, x, 77, 79–82, 86–88, 137

chloramphenicol (antibiotic), 85

Clothianidin (insecticide), 11–12, 15

Colony Collapse Disorder (CCD), viii–x,
9–10, 125

crossbreeding, 63, 70–71, 95–96. *See also*
Africanized honeybees; breeding;
Buckfast bees

cuckoo bumblebees, 123

dance, tail-wagging. *See* waggle dance

Darwin, Charles, 117–18

DDT (insecticide), 15

decision making, 52–53

Demeter bees/beekeepers, 34–36

disappearance (of bees). *See* Colony Collapse
Disorder

diseases. *See* American foulbrood; European
foulbrood; Isle of Wight disease; *Nosema*

dragonflies, 46

drone cutting, 107

drones, 25–27, 64, 67, 107, 120, 122

Dutch Gold Honey, 6

EFB. *See* European foulbrood

electric fields, 47

European foulbrood (EFB), 28–30

European honeybees. *See Apis mellifera*

eusocial bees, xi, 122

evolution, 14, 117–18, 120

eyes and vision, 40–41, 46–47

Fabini, Anne, 147–48

feral bees, 116–17

flight. *See* navigation and flight

formic acid. *See Varroa* mites: defense
strategies against

foulbrood (bacterial disease). *See* American
  foulbrood; European foulbrood
Friedmann, Günter, 36
Frisch, Karl von, 41, 43, 48, 124
fungicides. *See* agrochemicals

gardens, bee-friendly, 124
Gaucho (insecticide), 12
genetic diversity, lack of, xi, 69. *See also*
  breeding
Giurfa, Martin, 54
grafting, 65–67

Haefeker, Walter, 17, 71, 86, 89, 109
Hafenik, John, 11
Hamilton, Bill, 118
harvest shock, 92
Henry, Mickaël, 13
Heylighen, Francis, 56
"homeless" bees, 120
honey: contamination of, 5, 17–18, 85–86,
  110; medicinal purposes of, 88;
  production and harvesting of, viii, 33, 40,
  60, 74–76, 98–99; quality of, 8, 33, 85–86
honeycombs, 55, 75–76
honeydew honey, 74, 85
Hopfgartner, Peter, 145
horror movies, 98–99
hygiene, 32, 61

Imhoof, Markus, 15–16, 81, 85–86, 99; films
  of, 129–30

Imidacloprid (insecticide), 11, 15
inclusive fitness, 118
Indian bees, 19
industrialization of agriculture, 92–93
insecticides. *See* agrochemicals; *individual
  chemical names*
Isle of Wight disease, 69

Jaggi, Fred, 21–25, 27–31, 34, 135
Jeshel, Jörg, 144
jobs (of bees), 55–56
Johnson's organ, 47

Kaatz, Hans-Heinrich, 17
Kehrle, Karl (Brother Adam), 69–71
Kerr, Warwick Estevam, 95–99
"killer bees." *See* Africanized honeybees
kinship, 119–20
Koch, Ulrike, 136
Kramer, Ulrich, 23

lactic acid. *See Varroa* mites: defense
  strategies against
Langenbach, Jürgen, 13
leafcutter bees, 124
Lehnherr, Matthias, 36

MacIlvaine, Joe, 5, 18–19
macro filming (for documentary), 144–47.
  See also More Than Honey, making of
  documentary
magnetic fields, 47

manuka honey, 88

mating and reproduction, 25–27, 64–69, 73–74, 116, 121

McCallum, Brian, 16

Melissococcus plutonius (bacteria), 28

melittophily (pollination by bees), 40. *See also* pollination

Menzel, Randolf, 41, 43–45, 49, 52–54, 141–43

methylglyoxal, 88

Mexican migrant workers, 4

migratory beekeeping, 1–5, 74. *See also* beekeepers and beekeeping; pollination

Miller, John, 2–9, 19, 131–34

Miller, Nephi Ephraim (N.E.), 1

mining bees, 121

Monsanto (company), 16–18

More Than Honey, making of documentary, 80, 130–32, 134–48

Moritz, Robin, viii

myophily (pollination by flies), 40

navigation and flight, 13–14, 41–50

nectar, 31–33

neonicotinoids, 11–15

Nosema (disease), 8, 17

Nosema apis (fungus), 115–16

Oliver, Randy, 131, 138

orchard mason bees, 19

oxalic acid. *See Varroa* mites: defense strategies against

parasites. *See Apocephalus borealis; Nosema apis; Varroa* mites

pasture loss, 8, 121

Paucton, Jean, 89

Pavlov, Ivan, 43

pesticides. *See* agrochemicals; *individual chemical names*

pollen, 31–33, 81–82, 85–86, 123. *See also* pollination

pollination, viii, 2–4, 16, 31, 40, 80, 92, 122. *See also* migratory beekeeping

pollination, by humans, 79–82

Poncho (insecticide), 15

propolis, 60, 76–77

queen bees, 6, 25–27, 64–68, 74–75, 87, 120

Rademacher, Eva, 105

Reichholf, Josef, 91

reproduction. *See* mating and reproduction

research: American, 10–11, 14, 53; Australian, 114–16; European, 12–13, 15, 17, 43–44, 52–54, 70, 92–93; Latin American, 107–8

Ritter, Ruedi, 61

Robinson, Gene, 53

Rosenkranz, Peter, ix

royal jelly, 26–27, 55–56, 86–88

Schild, Elisabeth, 28

scouting bees, 53

Shi Wei, 81

Siberian bees, 84

Singer, Heidrun, 68, 135–36

Singer, Liane, 68, 136

Singer, Wolfgang, 136

Singer family, 68, 77, 135

smuggling, vii, 136

sociobiology, 118

Söder, Markus, 18

splitting (of bee colonies), 6, 26, 131

Steiner, Achim, ix

Steiner, Rudolph, 36

superbees, virus-resistant, 16

superorganisms, 120

symbolism, 36–37

Terry, Fred, 99–105, 111, 138–39

Thomas, Hans-Ulrich, 24

toxins. *See* agrochemicals

Unhoch, Nikolaus, 48

United Nations (UN), ix

urban beekeeping, 89, 91–92

Varroa mites *(Varroa destructor):* about, ix–x; in Australia, 113–14; defense strategies against, 8, 35, 84–85, 107, 109–11, 114–16; destruction caused by, 10, 92, 105; emergence of, 22, 83–84, 123; resistance/ tolerance to, 63, 82–83, 99–100, 107–8, 115–16

waggle dance, 14, 48–50

wasps *(Vespinae),* 39

wax, 55, 76

Whitehorn, Penelope, 13

wild bees, 104, 116, 120–24

winter bees, 56, 109

worker bees, 25–27, 64, 119

World Without Bees, A (Benjamin and McCallum), 16

Zhang Zhao, 137

Zhang Zhao Su, 82

Acknowledgments

For all their scientific advice and help with this book and the documentary *More Than Honey*, we thank Markus Imhoof's son-in-law, Prof. Boris Baer, his daughter, Dr. Barbara Baer-Imhoof, and the whole team at CIBER, University of Western Australia. Moreover, we thank Walter Haefeker, President of the European Professional Beekeepers Association and Thomas Radetzki, Melifera e.V., Germany.

DAVID SUZUKI INSTITUTE

The David Suzuki Institute is a non-profit organization founded in 2010 to stimulate debate and action on environmental issues. The Institute and the David Suzuki Foundation both work to advance awareness of environmental issues important to all Canadians.

We invite you to support the activities of the Institute. For more information please contact us at:

David Suzuki Institute
219 – 2211 West 4th Avenue
Vancouver, BC V6K 4S2
info@davidsuzukiinstitute.org
604-742-2899
www.davidsuzukiinstitute.org

Cheques can be made payable to The David Suzuki Institute.